Analyzing Mobile Apps Using Smart Assessment Methodology

Basheer Riskhan
Albukhary International University, Malaysia

Khalid Hussain
Albukhary International University, Malaysia

Halawati Abd Jalil Safuan
Albukhary International University, Malaysia

IGI Global
Scientific Publishing
Publishing Tomorrow's Research Today

Vice President of Editorial	Melissa Wagner
Director of Acquisitions	Mikaela Felty
Director of Book Development	Jocelynn Hessler
Production Manager	Mike Brehm
Cover Design	Jose Rosado

Published in the United States of America by
IGI Global Scientific Publishing
701 East Chocolate Avenue
Hershey, PA, 17033, USA
Tel: 717-533-8845
Fax: 717-533-7115
Website: https://www.igi-global.com E-mail: cust@igi-global.com

Library of Congress Cataloging-in-Publication Data

LCCN: 2024060477 (CIP Data Pending)
ISBN13: 979-8-3693-6925-8
Isbn13Softcover: 979-8-3373-1063-3
EISBN13: 979-8-3373-1064-0

British Cataloguing in Publication Data
A Cataloguing in Publication record for this book is available from the British Library.

Table of Contents

Detailed Table of Contents

Gamification has become a game-changer for mobile apps in education, making learning less like a chore and more like a challenge. This chapter dives into how different gamification techniques—like points, leaderboards, and badges—aren't just fun, but effective in keeping users engaged across various age groups, from kids to seniors. It takes me back to the first time I saw my niece breeze through her math homework on a gamified app, barely realizing she was learning. The psychology behind it is simple, give people a sense of progress, and they stick around. We'll look at some real-life examples—Duolingo, Fitbit, Kahoot—and how these apps have cracked the code. There's a bit of trial and error too, because what motivates one person might cause burnout in another. This chapter offers insights for app developers on what works, what doesn't, and how to push boundaries while keeping users engaged and motivated, all while making learning feel like less of a task and more like a triumph.

Device-free indoor localization is pivotal in smart environments, enabling precise positioning for applications like smart homes, healthcare, and IoT systems. This chapter presents an advanced localization framework optimized for mobile apps, integrating analytical geometry with machine learning techniques to overcome the challenges of dynamic indoor environments. Using Received Signal Strength

Indication (RSSI) data from strategically placed Wi-Fi modules, the system identifies precise locations without requiring individuals to carry devices. The methodology includes systematic data collection, preprocessing, and training of Random Forest models, ensuring robustness against environmental changes such as moving obstacles and varying occupancy. Experimental results highlight significant improvements in accuracy and responsiveness compared to conventional approaches, establishing the proposed method as a viable solution for real-world indoor localization scenarios. This work offers a scalable and efficient pathway for mobile applications in dynamic IoT ecosystems.

Mobile app security hinges on effective user authentication and authorization, yet we constantly see breaches due to weak safeguards. The chapter explores various methods to protect apps, from traditional password systems—often plagued by weak passwords and reuse—to advanced techniques like multi-factor and biometric authentication. But it's not just about getting people in; it's about making sure they access only what they should. Role-based and attribute-based access controls offer different ways to manage permissions, but if misconfigured, they can create loopholes. Throughout, smart tools play a big part in assessing these systems, catching flaws that developers may miss. Case studies drive home the risks, with real-world examples of fitness apps mishandling tokens or banking apps failing to secure biometric data. The chapter doesn't stop at the present; it also looks forward, examining how authentication and authorization will change with innovations like password-less access and decentralized identities.

Cloud computing is revolutionizing healthcare by improving service quality, efficiency, and affordability, while also addressing the evolving challenges in patient care. With the rapid digital transformation in healthcare, cloud-based systems offer new opportunities for data storage, patient monitoring, and healthcare management.

However, as more sensitive medical data moves to the cloud, security concerns regarding data privacy and third-party remote storage become critical issues. This research examines the potential of healthcare cloud computing, highlighting its benefits and challenges, including security and legal considerations. The paper introduces TDBNN, a deep learning-based system designed to enhance the security of cloud-stored healthcare data, ensuring better protection of patient privacy. This work provides insights into the future of cloud-based healthcare systems and their implications for patient-centered care.

Quantum computing brings about a new problem to traditional cryptographic algorithms; it violates the security of current encryption methods. This chapter is concerned with investigating two areas, quantum cryptography and post-quantum security measures, detailing their concepts, developments, and consequences. Quantum cryptography utilizes the principles of quantum mechanics and particularly makes use of superposition and entanglement. Post-quantum cryptography (PQC), on the other hand, is an approach to constructing cryptographic algorithms that will be immune to quantum attacks while at the same time keeping compatibility with classical methods. The chapter focuses on the aspects of integration between quantum cryptographical advancements and post-quantum algorithmic progressions and mutual distinctions have been explained. Possible shortcomings to scalability, implementation, and standardization issues of both approaches are discussed, as well as an examination of combined QKD and post-quantum algorithms architectures.

Mobile app monetization isn't just about revenue—it's about understanding what makes users stick around while developers keep their lights on. This chapter explores the strategies app developers use to earn while staying relevant. Whether it's enticing users with a freemium model, nudging them toward a subscription, or relying on good old-fashioned ads, each approach has its quirks. I remember downloading a so-called “free” app once, only to be bombarded with pop-ups asking for microtransactions to unlock anything fun. It’s a love-hate relationship, isn’t it? We dig into how factors like audience habits and app type shape these choices. Why do some apps thrive on in-app purchases while others struggle to sell a single upgrade? And let’s not forget about those success stories—Candy Crush and TikTok didn't happen by accident. But it's not all roses. Challenges like ad fatigue, app store policies, and balancing profitability with user trust keep developers on their toes. Looking ahead, the game is changing.

Attacks by malevolent software developers have increased in tandem with the growing popularity of cell phones. This study examines several techniques these assailants employ and recommends more effective ones. Android phones are especially susceptible to these attacks since network traffic analysis isn't taken into account sufficiently in today's detection techniques. In this article, a rule-based classifier—a cutting-edge technique—that can reliably identify more than 90% of hazardous traffic is introduced. Malicious apps are becoming more and more prevalent due to the expanding black market and the increasing number of applications that are accessible. To identify questionable activity, the suggested system automatically collects and examines applications from reputable retailers or unofficial marketplaces. The research provides a clear picture of how the platform has changed over time and how well various detection techniques work. It also compares several techniques for Android malware detection.

Banking sector has emerged into modern financial services and managing their transactions anywhere and anytime. This study explores the proposed frameworks based on the security of mobile banking systems focusing on multiple threats such as transaction fraud, phishing attacks, and unauthorized access. It focuses on the issues related to the user behavior that further leads into the vulnerabilities. Despite of focusing on the security level of mobile banking system, it is also important to focus on other measures like execution time consumption, storage, usage complexity and scalability. This study highlights the assessment of positive user experience towards the utilization of mobile banking applications. It is found that security enabled framework can provide best and effective results to the users if all other parameters are considered by researchers. Overall, this study findings are able to promote a reliable and authentic banking application in an emerging financial economy.

In this chapter, the methods to mobile app evaluation through smart assessment tools are explored. Traditional methods are not enough to capture user experience and app performance. A collection of smart methodologies that support data analytics and machine learning to enhance the evaluation process are presented here. The smart assessment tools make use of combination of real-time data and user centered design principles. With this combination, these smart assessment tools provide deeper perceptions into app usability and functionality. These methods further result in increased user satisfaction. There are many case studies where these methods are applied across many industries such as healthcare app evaluation, E-learning platforms and E-Commerce mobile apps.

The integration of Quantum Machine Learning (QML) into cloud-based security services represents a transformative shift in cybersecurity. By leveraging quantum computing's ability to process data exponentially faster than classical systems, QML enhances cloud security with real-time threat detection, anomaly recognition, and intrusion prevention. It introduces quantum-inspired algorithms, quantum versions of classical algorithms, and hybrid quantum-classical models, each offering unique advantages in processing speed and scalability. However, the rise of quantum computing also threatens conventional encryption methods like RSA and ECC, necessitating the development of quantum-resistant encryption techniques. Despite challenges posed by current quantum devices, QML's potential to revolutionize cloud security is immense, providing faster, more efficient, and robust protection against evolving cyber threats.

Preface

In an era where mobile applications shape our daily routines, from how we learn and communicate to how we bank and manage our health, there is an increasing need to assess these technologies not just for functionality, but for their impact, security, and long-term relevance. *Analyzing Mobile Apps Using Smart Assessment Methodology* brings together research from across disciplines and institutions to explore the tools, frameworks, and strategies necessary for evaluating mobile apps in smarter, more adaptive ways. Our aim with this volume is to provide a comprehensive reference for developers, researchers, and educators who want to understand not just how mobile apps work—but how they can work better for everyone.

The chapters included in this book reflect a rich diversity of topics—ranging from gamification in educational apps and device-free indoor localization, to security architectures for banking and quantum cryptography in cloud environments. At the heart of each chapter is a shared concern: how do we ensure that mobile apps are not only innovative but also secure, user-friendly, and ethically sound? The answer lies in smart assessment methodologies—data-driven, user-centered, and context-sensitive tools that allow us to move beyond surface-level metrics and into deeper, more meaningful evaluations of mobile application performance and trustworthiness.

As editors, we are honored to have worked with a global network of contributors who bring academic rigor, real-world insight, and visionary thinking to each chapter. The work presented here is both a snapshot of current advances and a blueprint for what comes next. By combining cutting-edge technologies like machine learning and quantum computing with human-centered approaches to design and evaluation, the chapters collectively highlight how mobile apps can continue to evolve in smarter, safer, and more purposeful directions.

CHAPTER OVERVIEW

Chapter 1: Gamification Techniques: Enhancing User Engagement With Mobile Apps for Learning

This chapter explores how gamification strategies—such as points, badges, and leaderboards—enhance engagement in educational mobile apps. Drawing on real-world examples like Duolingo and Kahoot, the authors examine how these techniques motivate users across age groups, highlight the psychological principles behind them, and offer practical insights for developers aiming to create learning experiences that are both effective and enjoyable.

Chapter 2: Optimizing Mobile Apps for Device-Free Indoor Localization in Dynamic Environments

Focusing on smart environments like IoT systems and healthcare facilities, this chapter presents a mobile app framework that uses analytical geometry and machine learning to enable device-free indoor localization. Leveraging RSSI data and Random Forest models, the method demonstrates improved accuracy in complex, dynamic environments, offering scalable solutions for location-aware mobile applications.

Chapter 3: Assessing User Authentication and Authorization in Mobile Apps

This chapter evaluates security mechanisms for mobile app access control, from traditional passwords to advanced biometric and multi-factor authentication. It highlights the role of smart tools in identifying flaws, examines access models like role- and attribute-based control, and discusses future trends including passwordless systems and decentralized identities.

Chapter 4: Revolutionizing Healthcare with Cloud Computing

The authors examine how cloud computing is reshaping healthcare delivery by enabling secure data storage, real-time patient monitoring, and system-wide efficiency. They address major challenges around data privacy and propose TDBNN, a deep learning model designed to safeguard patient information in cloud-based environments.

Chapter 5: Quantum Cryptography and Post-Quantum Security Strategies

This chapter delves into the emerging challenges and solutions in cryptography brought on by quantum computing. It compares quantum cryptographic techniques with post-quantum algorithms, discusses integration approaches, and analyzes the scalability and standardization concerns that arise in building future-ready security systems.

Chapter 6: Evaluating Mobile App Monetization Strategies

Exploring various revenue models such as freemium, subscriptions, and ads, this chapter analyzes how developers balance profitability with user experience. Through case studies and user behavior insights, it highlights both the successes and challenges in mobile app monetization, including issues like ad fatigue and evolving app store policies.

Chapter 7: Detecting and Analyzing Malware in Android Mobile Devices

This chapter presents a rule-based classifier for detecting Android malware through network traffic analysis, addressing gaps in current detection methods. It outlines how malicious apps spread via marketplaces, compares detection techniques, and emphasizes the growing need for intelligent, real-time monitoring systems.

Chapter 8: Smart Performance Analysis of Security-Enabled Mobile Apps for Banking

Focusing on mobile banking apps, this chapter evaluates frameworks that address both security threats and performance factors like execution time, scalability, and user complexity. The findings suggest that an integrated approach to security and usability can enhance trust and functionality in financial mobile applications.

Chapter 9: Innovative Methods for Mobile App Evaluation Using Smart Assessment Tools

This chapter introduces smart assessment tools that combine real-time data and user-centered design to evaluate mobile app usability and performance. With applications across healthcare, e-learning, and e-commerce, these tools enhance user satisfaction by providing deeper insights than traditional evaluation methods.

Chapter 10: Quantum Machine Learning in Cloud-Based Security Services

Exploring the fusion of quantum computing and machine learning, this chapter investigates how Quantum Machine Learning (QML) can bolster cloud security through rapid threat detection and anomaly recognition. It also addresses the limitations of current quantum devices and the urgent need for quantum-resistant encryption methods.

We believe that this book arrives at a crucial time, as mobile applications become more central to everyday life and digital ecosystems become more complex and interconnected. The tools and methods for assessing apps must rise to meet these challenges. This volume is a step in that direction, offering not only academic findings but also practical guidance for implementation in real-world scenarios. The editors and contributors share a collective belief in the transformative potential of smart assessment, particularly when combined with ethical design and cross-sector collaboration.

It is our hope that this work sparks continued dialogue and innovation. The smart assessment methodologies discussed here are not the final word, but rather a foundation upon which future research and development can build. We encourage readers to adapt, extend, and challenge the ideas presented—to consider not only what apps do, but how and why they do it. As technologies like AI, quantum computing, and immersive interfaces continue to reshape mobile platforms, ongoing critical evaluation will be essential to ensuring they remain effective, inclusive, and secure.

Finally, we wish to extend our deepest gratitude to the contributing authors for their expertise and commitment, and to our academic institution, Albukhary International University, for supporting this project. To all readers, whether you are developers, researchers, educators, or policy makers: thank you for joining us on this exploration of mobile app assessment. May the insights within these chapters guide your work, spark your curiosity, and inspire smarter solutions for the mobile-first world ahead.

Chapter 1
Gamification Techniques: Enhancing User Engagement With Mobile Apps for Learning

Siva Raja Sindiramutty
https://orcid.org/0009-0006-0310-8721
Taylor's University, Malaysia

Noor Zaman Jhanjhi
https://orcid.org/0000-0001-8116-4733
Taylor's University, Malaysia

ABSTRACT

Gamification has become a game-changer for mobile apps in education, making learning less like a chore and more like a challenge. This chapter dives into how different gamification techniques—like points, leaderboards, and badges—aren't just fun, but effective in keeping users engaged across various age groups, from kids to seniors. It takes me back to the first time I saw my niece breeze through her math homework on a gamified app, barely realizing she was learning. The psychology behind it is simple, give people a sense of progress, and they stick around. We'll look at some real-life examples—Duolingo, Fitbit, Kahoot—and how these apps have cracked the code. There's a bit of trial and error too, because what motivates one person might cause burnout in another. This chapter offers insights for app developers on what works, what doesn't, and how to push boundaries while keeping users engaged and motivated, all while making learning feel like less of a task and more like a triumph.

DOI: 10.4018/979-8-3693-6925-8.ch001

1. INTRODUCTION

Definition of Gamification

Gamification is the use of game-like elements in non-gaming settings to boost engagement and motivation (Sanchez et al., 2020; Christopoulos & Mystakidis, 2023; Ananna et al., 2023). It's interesting how this concept has found its way into various industries, from education to marketing. Imagine being in a classroom where, instead of just sitting through lectures, you earn points for participation or completing tasks—suddenly, the environment feels a bit more fun, right? That's the power of gamification. It's not about playing games, but rather borrowing elements like leaderboards, achievements, and rewards to encourage desired behaviours (Mekler et al., 2013).

In education, gamification has proven to be effective by making learning less of a chore (Orji et al., 2023). Students are more likely to stay motivated when their progress is rewarded in ways they enjoy (Trinh et al., 2023; Gopi et al., 2021). It's like how we stick with mobile apps that give us badges for hitting milestones. I remember trying to learn a new language through a gamified app and being surprisingly hooked by the tiny rewards I kept earning, even though I wasn't particularly interested in the language itself. That's the magic of gamification. In marketing, companies use gamification to build customer loyalty (Singh & Milan, 2024; Gouda et al., 2022). Loyalty programs are a great example, where customers earn points for purchases, much like players collecting tokens in a game (Vaitone, 2025). It makes the experience feel more interactive and less transactional. The success of gamification lies in its ability to tap into basic human desires—competition, achievement, and status (Sharma et al., 2023). By turning ordinary tasks into opportunities for fun and recognition, gamification transforms everyday experiences. Figure 1 shows gamification purpose.

Figure 1. Gamification Purpose (Harbinger Group, 2024)

Importance of user engagement in mobile apps for learning

Getting people hooked on learning apps takes more than just pretty interfaces and easy-to-use features. It's all about keeping users engaged, and that can make or break the success of educational mobile apps. I remember downloading a language-learning app, using it non-stop for a week, and then just... forgetting about it. Why? Because it didn't keep me interested. That's the thing—if users don't feel involved, they're likely to quit. Studies show that interactive features like quizzes, push notifications, and rewards increase engagement, leading to better learning (Wang & Sari, 2023). Gamification also plays a huge role (Denden et al., 2022; Humayun et al., 2022). When apps use game-like elements—think leaderboards or badges—they tap into the human need for achievement, which encourages consistent use (Gil-Aciron, 2022; Jhanjhi et al., 2021). On the flip side, apps that feel too passive or repetitive quickly lose their charm. Even social features, like being able to share progress with friends, help keep people invested in their learning journey (Scanlon et al., 2024).

There's another angle to this. Personalized content makes users feel seen, increasing their motivation to return to the app (Dastane et al., 2023; Kumar et al., 2021). If an app recommends lessons based on a user's weak spots or goals, it feels tailored just for them, and that personal touch makes all the difference. Without these interactive and personalized elements, mobile learning apps might struggle to hold users' attention in today's fast-paced digital world (Nurcahyo, 2024). So, designing an engaging experience isn't just a bonus—it's essential.

Learning through Gamification in Digital Curriculum

Gamification in digital learning has sparked interest because of its potential to make education more engaging (Hong et al., 2024; Lim et al., 2019). Imagine trying to learn a tough subject like math, but instead of endless equations, you're playing a game that rewards you for solving problems. It's not just about making learning fun; it's about keeping students hooked, just like they'd be with a video game. I've seen firsthand how my younger cousin, who hated history, suddenly became fascinated when it was presented in a game format. He raced to finish quests, earning badges, all while picking up historical facts without even realizing it. Research supports this, showing that gamification helps improve motivation and engagement (Bachoo & Nagowah, 2024).

In a broader sense, gamification isn't just about points and rewards—it can also offer instant feedback, which can speed up the learning process (Ibisu, 2024). A friend of mine told me about his son's school that uses gamified apps for homework, and he says it's made a noticeable difference in his son's confidence (Li et al., 2024; Azam, Dulloo, Majeed, Wan, Xin, & Sindiramutty, 2023). It's fascinating to see how digital platforms can transform learning from something students dread into something they actively look forward to (Khan, 2024; (Nayyar et al., 2021). Beyond that, the interactive nature of gamification helps with retention, as students are more likely to remember what they've learned when it's tied to an experience (Jansukpum et al., 2024).

And while it might seem like just another trend, this approach seems to be gaining momentum in education, offering a fresh way to teach even the most challenging topics. It just might be the future of how we all learn.

Overview of the chapter's objectives

The chapter provides a fresh perspective on how gamification can enhance user engagement in mobile learning apps. By integrating elements like points, leaderboards, and badges, these apps motivate users to keep participating, appealing to our natural desire for achievement. The chapter explores real-life applications, showcasing how successful apps such as Duolingo and Fitbit have harnessed gamification to make learning engaging and rewarding. For instance, the use of leaderboards fuels healthy competition, pushing users to reach higher scores, while badges provide instant recognition, sparking the drive for continuous progress. The narrative incorporates examples from diverse age groups, showing how kids to seniors are drawn to learning apps due to these interactive features. Moreover, the chapter touches on the challenges gamification faces, such as balancing motivation with potential burnout when the same techniques don't appeal to everyone. With emerging technologies like AR and VR, the potential for gamified learning seems boundless, making it an exciting area for further exploration and innovation.

Chapter Organization

In the first section, we will cover an insight into the introduction gamification and importance of user engagement in mobile apps for learning. In section 2 we go down statistics of user time with mobile screen for different age groups. In section 3 we the psychology behind gamification followed by key gamification techniques such as point system, leaderboards progress tracking in section 4. Section 5 delves case studies of successful gamification in mobile apps. Section 6 will explain the challenges and considerations. in final section will discuss future trends in gamification.

Chapter Contribution

This chapter makes a valuable contribution by shedding light on how gamification techniques can transform user engagement in mobile learning applications. By analyzing the impact of game-based elements like points, badges, and leaderboards, it shows how these strategies not only draw users in but keep them motivated over time. It offers practical insights into the psychology behind these techniques, focusing on how elements like rewards and social features can satisfy users' needs for accomplishment, social connection, and autonomy

Through real-world examples such as Duolingo and Fitbit, the chapter demonstrates how gamification has succeeded in creating engaging learning experiences across various demographics, from children to adults By also addressing challenges like user fatigue and the potential for diminishing returns, the chapter provides a balanced view, guiding developers in crafting user-centered designs that are both effective and sustainable Furthermore, the chapter introduces emerging trends like AR and VR as avenues for expanding gamification's reach, hinting at a future where digital learning feels more immersive and enjoyable than ever. This insight aims to inspire educators and app developers to explore innovative ways of making learning accessible, rewarding, and fun for all users.

2. UNDERSTANDING USER ENGAGEMENT

Definition and key components of user engagement

User engagement is the interaction between people and digital platforms, whether it's a website, app, or social media (Jha & Verma, 2023; Shah et al., 2022). It's that feeling you get when something grabs your attention and keeps you coming back. The key components that drive engagement include attention, emotion, participation, and feedback. When users feel emotionally connected, they're more likely to stick around and interact with the content (Schluchter et al., 2024). Think about how a funny meme or an engaging article can spark some reaction in you — that's the emotional pull.

Attention plays a huge role, too. In today's fast-paced digital world, you can lose people's attention in seconds. Platforms that keep things interesting, with a blend of visuals and easy navigation, tend to succeed in holding users longer (Siricharoen, 2023; Azam, Dulloo, Majeed, Wan, Xin, Tajwar, et al., 2023). Participation is another crucial element. When users can contribute or feel involved, like leaving comments or joining a community discussion, their engagement deepens. It's like being part of a conversation rather than just listening in.

Finally, feedback loops matter. Users want to feel heard. Responding to comments or improving features based on feedback shows that their participation counts (Demszky et al., 2023). All these components combine to create a memorable and engaging user experience, making people return again and again.

Statistics of User time with mobile screen for different age groups

People today seem glued to their phones, but the amount of screen time varies wildly depending on age. Teens, for example, are reported to spend an average of 7 hours per day on their devices (Zen et al., 2024; Sharma et al., 2021). It's easy to imagine, given how often you see them staring at screens during school breaks or even while crossing the street. Interestingly, younger kids aged 8-12 spend around 4.5 hours daily on devices (Pekpak & Altunalan, 2024; Ying et al., 2024), but my niece, who's 9, easily goes beyond that when playing her favourite mobile game. Adults, on the other hand, aren't exactly innocent either. Research suggests those in their 30s and 40s clock in about 3-4 hours of mobile screen time each day (Medellin & Saluy, 2020; Azam, Tajwar, Mayhialagan, Davis, Yik, Ali, et al., 2023; Singhal et al., 2020). I often catch myself checking emails and scrolling social media way past that estimate. Meanwhile, people aged 55 and older spend closer to 2-3 hours a day on their phones. I've noticed my parents, both in their 60s, use their phones mainly for news or video calls with family.

The real question is what all this screen time is doing to our attention spans and daily lives. Excessive use, especially among younger users, has been linked to sleep problems and issues with attention (Ramadani, 2024); Weiqi et al., 2024). I've personally found it harder to focus after a day of constant phone-checking. The implications are significant, impacting both mental and physical health. Yet, as screens become more integral to our lives, it's hard to imagine going back to how things were before. Still, a little moderation might not be a bad idea. Figure 2 shows key components of user engagement.

Figure 2. Key Components of User Engagement

Interactivity	How often and actively users interact with the platform, such as clicks, swipes, or comments. Social media likes, sharing content, leaving comments.
Frequency	The regularity of user visits or interactions with the platform. Daily app usage, weekly logins, or monthly active users.
Satisfaction	The perceived value and satisfaction users feel from their experience. Positive feedback, user ratings, net promoter score (NPS).
Retention	The likelihood of users returning over time. Returning user rate, customer lifetime value (CL)
Conversion	The rate at which engagement translates into actions like purchases or subscriptions. Signup rates, purchase frequency.

The role of engagement in app success for learning

When it comes to learning apps, engagement is often what makes or breaks them. I remember trying out a popular language app, and while its design was polished, the daily reminders and fun streaks kept me hooked. This level of interaction creates a habit, making it hard to skip even a single day of learning. According to a study, users are more likely to continue using an app that feels rewarding (Huang et al.,

2024; Sindiramutty, 2024). It's like playing a game—if you're engaged, you keep going, and before you know it, you've covered a ton of material. That's how these apps work. They tap into the psychology of motivation, keeping you curious.

The real trick is when apps combine learning with challenges, badges, and social sharing. My cousin, who struggles with staying focused, suddenly found himself eager to complete math exercises just to earn points and beat his friends. Research shows that features like leaderboards increase competition and motivation among learners (Chen & Huang, 2024; Sheng et al., 2024). This not only improves their overall experience but also keeps them coming back.

Of course, it's not just about points. Engagement can mean having intuitive, user-friendly designs that don't overwhelm. A cluttered app? Forget it—I won't last five minutes. But one that lets me smoothly navigate between lessons? I'm all in. As Kuh (2009) highlights, easy-to-use interfaces keep learners focused, rather than frustrated (McKay, 2013). Engagement is the secret sauce that makes these apps stick and ultimately leads to better learning outcomes (Zosh et al., 2016).

Statistics on user retention and engagement in mobile apps

When I first started using mobile apps, I never really thought about how much effort goes into keeping users like me coming back. However, it turns out that user retention and engagement are major concerns for app developers. I remember reading somewhere that the average retention rate for mobile apps drops to just 25% after the first day of use (Audestad & Telematikk, 2014; Linqiang et al., 2024). That's shocking! Imagine developing an app, only to see most users abandon it after the first try.

A big part of this problem comes down to engagement. Apps that encourage interaction, like social media or gaming apps, tend to keep users around much longer (Anvari et al., 2024; Azam, Tan, Pin, Syahmi, Qian, Jingyan, et al., 2023). I've seen it myself; the more interactive and enjoyable an app feels, the more likely I am to return. On the flip side, apps that bombard users with notifications or overly complicated features tend to lose them quickly (Temel, 2024).

Interestingly, data shows that personalization can increase retention rates. If an app feels like it "knows" me by showing content based on my preferences, I'm more inclined to stick around. But even then, developers face challenges. In the end, app developers have to strike a balance between engaging features and user satisfaction or risk losing their audience faster than they can attract them.

3. THE PSYCHOLOGY BEHIND GAMIFICATION

Explanation of motivation and behavior theories

Gamification taps into human psychology to inspire motivation and influence behavior, with Self-Determination Theory (SDT) offering a helpful lens for understanding this. SDT proposes that people are driven by a need for autonomy, competence, and relatedness (Landers et al., 2014). Gamified experiences fulfil these needs by offering a sense of control (autonomy), clear challenges (competence), and social connections through interaction (relatedness) (Xi & Hamari, 2019; Hussain et al., 2024). When I first started using a fitness app with game-like elements, I found myself far more motivated to stick to my exercise routine simply because I could "level up" and earn rewards. It felt less like a chore and more like I was mastering something fun hitting the competence mark.

Behavioural theories also weigh in. The concept of operant conditioning explains how rewards like badges and points reinforce positive behaviour (Troussas et al., 2017; Jun et al., 2024). This taps into a kind of Pavlovian response where the brain anticipates the reward and encourages us to continue (Burghoorn et al., 2024). My friend swore off reading for years, but then she downloaded a reading app that rewarded her with achievements. Before she knew it, she was reading daily, all for the dopamine hit of earning a badge.

Interestingly, the social aspect, especially in multiplayer games, aligns with the theory of social influence, where people modify their behaviour based on social interaction (Gong et al., 2023). In competitive or cooperative environments, players often feel encouraged to push themselves further just to keep up or stand out in their peer group (Moller et al., 2024; Kiyani et al., 2024). This mirrors my own experience, where I felt strangely compelled to walk more steps in a day than my friends just to keep my name at the top of the leaderboard. Figure 3 shows the psychology behind gamification.

Figure 3. The Psychology Behind Gamification

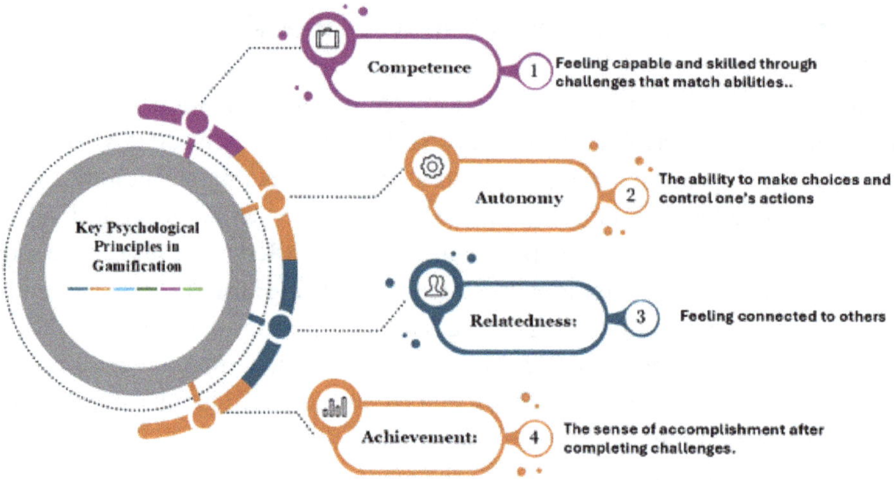

The impact of gamification on user behaviour

Gamification uses game-like elements in non-game contexts to motivate and engage users (Tongpaeng et al., 2024). It taps into human psychology, influencing behaviour and decision-making. Think about how earning points in a fitness app or badges in an online course makes you feel accomplished. These simple rewards trigger the release of dopamine, a chemical associated with pleasure, making you want to engage more (Hunt et al., 2024; Manchuri et al., 2024). The excitement of competition also plays a role. Leaderboards can spark a healthy rivalry, encouraging users to push themselves further.

In my own experience, using a language-learning app made the process much more enjoyable. I looked forward to earning daily streaks and badges, which kept me coming back for more. Research shows that this kind of reward system increases engagement and persistence (Pombo & Santos, 2023). By making tasks feel like a game, people are more likely to stick with them.

Another factor is social interaction. Many gamified platforms allow users to share achievements and compete with friends (Zhang et al., 2023; Ravichandran et al., 2024). This sense of community creates a supportive environment that drives participation. When you know your friends are watching your progress, it can motivate you to keep going.

Ultimately, gamification can significantly change how users behave. By understanding the psychological triggers at play, designers can create experiences that keep people engaged and motivated. The blend of competition, rewards, and social connections fosters an environment where users are more likely to act.

Examples of psychological triggers used in gamification

Gamification uses psychological triggers to motivate people in various settings, from education to business (Sotos-Martínez et al., 2022). It taps into our instincts, making tasks feel more like games. For instance, **leaderboards** create a competitive environment, encouraging users to strive for higher rankings (Cao et al., 2022; Seng et al., 2024). This competition often spurs individuals to work harder, aiming to outperform peers. Another powerful trigger is **achievement badges**. When users earn these, it boosts their self-esteem and encourages them to engage more with the task (Baidya et al., 2024). It's like collecting stickers in school—everyone loves a tangible reward for effort!

Progress bars are also common; they show how far you've come, igniting that urge to complete a task. Seeing that visual representation of progress gives a sense of accomplishment, pushing individuals to finish what they started. Then there's the trigger of **social interaction**. Many gamified systems incorporate elements where users can collaborate or share achievements with others, enhancing motivation through community support (Chan et al., 2023; Sindiramutty, Prabagaran, Jhanjhi, Murugesan, Brohi, & Masud, 2024). You might think of it as cheering for your favourite team—having others celebrate your success feels great!

Finally, the aspect of **random rewards**—like surprise gifts or points—can keep people engaged by introducing an element of unpredictability (Brown et al., 2023). This unpredictability makes users return for more, as the next win could be just around the corner. These psychological triggers illustrate how gamification effectively taps into our behaviours and emotions, transforming mundane tasks into engaging activities.

Comparison of Gamification learning results with conventional learning

Gamification has changed how we approach learning, tapping into the psychology of motivation to boost engagement and retention (Mahmoudi-Dehaki & Nasr-Esfahani, 2024). Traditional learning often feels tedious, leading to disengagement. On the other hand, gamified learning transforms lessons into interactive experiences, making students more eager to participate (Cahya & Indiana, 2023). This approach is like adding a sprinkle of excitement to a mundane task—think of it as turning a math lesson into a thrilling quest. Studies show that learners in gamified environments often perform better than those in conventional settings (Chapman et al., 2023). For instance, a study by Bolat and Taş (2022) found that gamification can significantly enhance motivation and achievement in educational contexts.

Moreover, a meta-analysis by Hellín et al. (2023) indicated that elements like points, badges, and leaderboards could foster a sense of competition and community among learners, driving better results. In one real-life case, a university integrated game mechanics into their curriculum, which led to a noticeable increase in student participation and grades. Those who engaged with gamified content scored higher on assessments compared to their peers (Cigdem et al., 2024).

It's essential to consider that while gamification can provide immediate benefits, it might not suit everyone. Some learners may thrive in structured, traditional environments. Ultimately, it's about finding the right balance. Gamification offers a fun twist that might just reignite the spark for learning, making it not just about scores and grades, but about the joy of discovering new knowledge.

4. KEY GAMIFICATION TECHNIQUES

Points Systems

Gamification often uses point systems to boost user engagement and motivation in various settings, like education and business (Bizzi, 2023; Sindiramutty, Prabagaran, Jhanjhi, Ghazanfar, Malik, & Soomro, 2024). These systems reward participants for completing tasks or achieving goals, turning mundane activities into fun challenges. I remember when my friend started using a fitness app that awarded points for every workout. It made exercise feel like a game, and he was much more consistent in reaching his fitness goals.

Points systems serve several purposes. First, they provide immediate feedback. When users see points accumulate, it reinforces their efforts and encourages continued participation (L. Zhang et al., 2023). Secondly, they can create competition. Users often compare scores, fostering a sense of community and engagement. This competitive element can motivate individuals to perform better, whether in learning or workplace productivity (Oke et al., 2023)

Moreover, point systems can help set clear objectives. Participants know exactly what actions will earn them points, making the path to success straightforward. This clarity can be particularly beneficial in educational settings, where students might struggle with self-directed learning (Landers, 2014). Ultimately, point systems are more than just a way to track progress; they add excitement, encourage interaction, and make achieving goals feel rewarding.

Examples of apps using points systems

Gamification, especially through point systems, can change how we interact with apps by making tasks feel more like a game. For instance, fitness apps like **MyFitnessPal** reward users with points for logging meals and workouts, pushing them to reach their health goals (Morisio, n.d.). Similarly, **Duolingo** uses a points system to motivate language learners. Users earn points for each lesson completed, fostering a competitive spirit as they aim for higher scores (Shortt et al., 2021).

In the realm of finance, apps like **Mint** utilize points to encourage budgeting habits (Mazlan, 2024; ; Sindiramutty, Prabagaran, Akbar, Hussain, & Malik, 2024). Users can earn points by sticking to their spending goals, which makes managing money a little more fun. Even in the world of shopping, apps like **Sephora** give points for purchases, creating a loyalty system that keeps customers coming back for more (Herbach, 2024).

It's interesting to see how these points systems tap into our desire for achievement and recognition. They don't just keep us engaged; they foster communities where users share their accomplishments and strategies. That's something I've noticed firsthand when using **Fitbit**—seeing friends' points motivates me to move more. In a world where achieving daily goals can sometimes feel like a grind, point systems add a layer of excitement that can turn the mundane into something extraordinary. Figure 4 shows key gamification techniques.

Figure 4. Key Gamification Techniques

Leaderboards

Role in Fostering Competition

Leaderboards are a popular gamification technique used to foster competition by ranking participants based on their performance (Do et al., 2024). In my experience, seeing my name lower on a leaderboard always pushed me to try harder, even if I didn't fully understand why it mattered so much. This simple ranking system taps into our natural desire to win and be recognized, triggering motivation to outdo others (Gao et al., 2024; Sindiramutty, Prabagaran, Jhanjhi, Murugesan, Malik, & Hussain, 2024). But it's not just about being number one. Even those in the middle of the pack often feel driven to climb higher, as leaderboards highlight who's ahead and who's catching up.

Once, I was in a coding challenge, and even though I wasn't leading, just knowing someone close to my level was right above me pushed me to keep improving. This competitive aspect not only encourages individual achievement but can also strengthen community bonds, as participants push each other forward (Modarelli & Rainero, 2024). In educational settings, leaderboards have been found to increase engagement, though some argue they may demotivate those at the bottom (Zainuddin et al., 2024). But in most cases, their role in creating a more engaged and competitive environment is undeniable. It's funny how just seeing your name move up or down on a digital board can feel like such a personal triumph—or defeat.

Case studies of successful implementations

One memorable example of leaderboards being used effectively is Duolingo, a language-learning app. I remember when I first started using it, seeing my rank compared to others gave me an extra push to practice daily. Studies have shown leaderboards can lead to more consistent participation because users feel a sense of competition and accomplishment (Slamet et al., 2024). Another case is Fitocracy, a fitness app that uses leaderboards to track user workouts, pushing participants to log more activity as they strive for higher ranks (Eklund et al., 2018).

Leaderboards aren't limited to apps. In workplace settings, Salesforce implemented a leaderboard system for sales teams. Salespeople who saw their rank climb were more motivated to outperform their peers (Ammattikorkeakoulu, 2015; Sindiramutty, Tan, Shah, et al., 2024). Even in education, platforms like Classcraft use leaderboards to encourage students to complete tasks, helping improve learning outcomes (Zainuddin et al., 2024b)

Of course, leaderboards aren't perfect. They can sometimes demotivate those who struggle to keep up. I've seen people give up when they feel too far behind. However, with proper design and support, leaderboards can drive participation, create a friendly sense of competition, and build a community of achievers (Nicholson, 2015).

Badges and Achievements

Significance in user motivation

Gamification has become a popular strategy to keep users motivated, especially through the use of badges and achievements (Mohammed et al., 2024). These small rewards act as markers of progress, encouraging people to keep pushing forward. I remember when I first earned a badge in a fitness app, and it instantly gave me that extra push to work out more consistently. These badges offer instant recognition, which can fuel the desire to achieve even more (Chiu et al., 2024). People love feeling acknowledged, and earning an achievement – no matter how small – taps into this basic human need for recognition (Rusin & Szandała, 2024; Sindiramutty et al., 2024). A friend of mine, for instance, kept using a language learning app just because she wanted to collect all the badges. It seems a little silly, but it worked! Research also supports this, showing that badges enhance engagement and can even change user behaviour (Malik et al., 2024). Achievements, meanwhile, create a sense of accomplishment, boosting users' confidence and willingness to continue. It's fascinating how something as simple as a virtual badge can have such a strong motivational effect. From personal experiences to broader studies, it's clear that badges and achievements play a significant role in keeping users engaged in digital platforms.

Examples of apps that reward achievements

Gamification has become popular in many apps, using badges and achievements to keep users engaged (Petrou et al., 2024). I've always been hooked by how apps reward milestones—earning a badge just makes me feel good, even if it's for something small. One of the standout examples is Duolingo, where users get badges for language progress, making me keep coming back to beat my streaks (Neuschafer, 2023; Waheed et al., 2024). Similarly, in fitness apps like Fitbit, badges are given for hitting step goals or completing challenges with friends, which taps into that competitive spirit. I've seen people push their limits just to show off these badges! LinkedIn has its version too. It uses achievement badges for profile strength or skill endorsements, which makes professionals strive for more connections and skills (Arroteia, 2024). Even in gaming, platforms like Xbox reward players with achievement badges for reaching in-game goals. It's interesting how something so simple can drive so much engagement—like it makes the digital experience feel a bit more personal. Apps like Habitica turn daily habits into a game, rewarding tasks completed with badges and giving users that same sense of accomplishment (Hall-Johnson & Hall-Johnson, 2023). Isn't it fascinating how little digital tokens can push us to do more?

5. CASE STUDIES OF SUCCESSFUL GAMIFICATION IN MOBILE APPS

Analysis of mobile apps that effectively use gamification (e.g., Duolingo, Fitbit, and Kahoot)

Gamification in mobile apps has taken off in recent years, creating engaging experiences that keep users coming back for more (J. Zhang, 2023; Wen et al., 2023). Duolingo is a great example. I find it amusing how those cute little owl reminders make me feel guilty about missing a lesson! They use points, streaks, and challenges to keep learning fun and, honestly, a bit addictive. Then there's Fitbit—tracking steps never felt rewarding until you started earning badges for hitting milestones. It's like a fitness game, and the friendly competition with friends always adds an extra push. Kahoot, on the other hand, takes quizzes to a whole new level with its game-like approach (Ebadi et al., 2021). I remember using it during meetings; suddenly, everyone's competitive side came out, and no one even cared that it was still "work". These apps succeed by turning ordinary tasks into enjoyable, repeatable actions. Gamification makes users feel like they're achieving something beyond the app's actual goal (Kunkel et al., 2023). The blend of instant rewards and longer-term goals hooks people, whether it's learning a new language, staying active, or taking a quiz. They've shown how powerful gamification can be in transforming user experiences across industries. Figure 5 shows successful gamification in mobile apps.

Figure 5. Successful Gamification in Mobile Apps

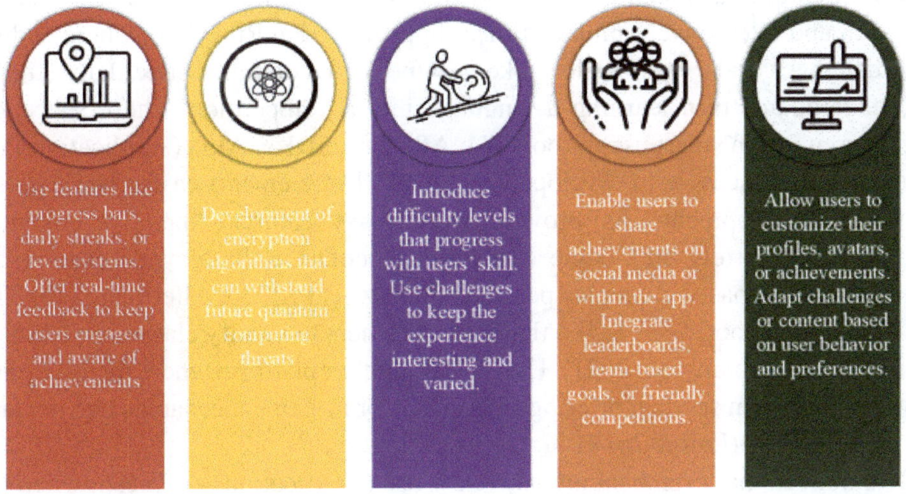

Steps for integrating gamification techniques

Gamification in mobile apps has taken off in ways that few could have predicted, turning mundane tasks into something a bit more exciting (Vashisht, 2023). I remember trying an app to track water intake, and it had badges for simply staying hydrated—I stayed hydrated just to win a virtual trophy! One app that stands out is Duolingo, which integrates point systems and leaderboards to push users to practice more (Kastelli & Takács, 2023). It's crazy how something as small as earning "XP" can make language learning feel like a game. And it's not just language learning; apps like Nike Run Club use challenges and rewards to motivate users to stay active (Nigam & Chanda, 2024). Fitbit, too, encourages users by giving virtual badges when they reach certain step goals, adding that sense of accomplishment (Ong & Samson, 2024; Alex et al., 2022). In building these systems, it helps to break down tasks into clear goals. Adding small achievements along the way makes users feel like they're constantly progressing. Even productivity apps, like Habitica, have managed to turn daily routines into fun quests (Ong & Samson, 2024). The secret seems to be in rewarding users frequently and keeping them engaged with some playful competitiveness. It's amazing how adding a little fun can make us all a bit more productive—or hydrated!

Tools and platforms for gamification (e.g., APIs, SDKs)

Gamification has taken mobile apps to a new level, making user experiences more fun and interactive. Think of fitness apps like Nike Run Club, where badges and challenges make running feel like a game, or Duolingo, where daily streaks and leaderboards turn learning a language into a competition. These successes largely come from smart use of tools like APIs and SDKs. APIs (Application Programming Interfaces) help developers add game-like elements easily, while SDKs (Software Development Kits) provide ready-to-use features. Apps like Pokémon Go, for instance, rely on Google Maps API to enhance the player's experience, making location-based gameplay possible (Juhász et al., 2020; Alferidah & Jhanjhi, 2020). Also, Kahoot! Uses SDKs that allow teachers to gamify classroom learning (Sukmana & Rosmansyah, 2021). Unity3D is another platform where many gaming apps are built from scratch, offering rich SDKs for in-game mechanics like rewards and leaderboards (Laine, 2018).

I've noticed that even banking apps now use gamification — I mean, when my banking app rewards me for spending less, it kind of makes me want to save more! This blend of tools, from APIs to SDKs, shows how apps across different industries are successfully pulling users in, making mundane tasks more exciting.

6. CHALLENGES AND CONSIDERATIONS

Potential pitfalls of gamification (e.g., over-justification, user burnout)

Gamification has gained popularity in learning and workplace settings, but it's not without its challenges (Oravec, 2015; Alkinani et al., 2021). One common issue is over-justification. People might start engaging in activities only for rewards, losing interest in the task itself. I've seen this happen firsthand when I introduced gamified tasks to a group of students. At first, they were excited about earning points. But soon, they started caring more about the rewards than learning. Another problem is burnout. Constantly striving for badges and achievements can wear people out. I've experienced this with apps that bombard users with notifications, pushing them to complete daily goals. It becomes exhausting, and soon, the excitement fades.

Gamification also risks alienating users if the system feels unfair or too difficult (Mellado, 2024). It's frustrating when you're working hard but the rewards seem unreachable. There's also the problem of gamification being one-size-fits-all. Some people thrive on competition, but for others, it's a source of stress. Finally, there's the issue of superficial engagement. If rewards are too easy, they lose their value.

People may engage only for the sake of ticking boxes, not because they find the experience meaningful.

Ethical considerations in gamification

Gamification has quickly become a popular tool for engaging users, but it also brings ethical questions that deserve attention (Karpouzis, 2024; Babbar et al., 2021). When I first encountered it, I thought gamifying things like education was genius. Who doesn't want learning to be fun? But then, I started thinking about how these systems manipulate behaviour. Gamification can sometimes cross into exploiting users' instincts for rewards, creating addictive patterns or pressuring people to compete in unhealthy ways. If companies don't use it responsibly, they might end up nudging people to spend more time or money than they would naturally want to, especially on online platforms and apps.

Take those fitness apps that guilt-trip you with notifications about missed goals. It's a subtle pressure, but it's real. The risks also extend to privacy. A lot of gamified systems collect data about users' habits, which can be sold or used in ways people might not be aware of. Some critics argue that this takes advantage of people's trust, especially when users are unaware of how their data is being handled. It's a fine line between motivation and manipulation, and one that developers and organizations need to walk carefully if they want to maintain ethical practices. Figure 6 shows challenges in gamification. Figure 6 shows challenges in gamification.

Figure 6. Challenges in Gamification.

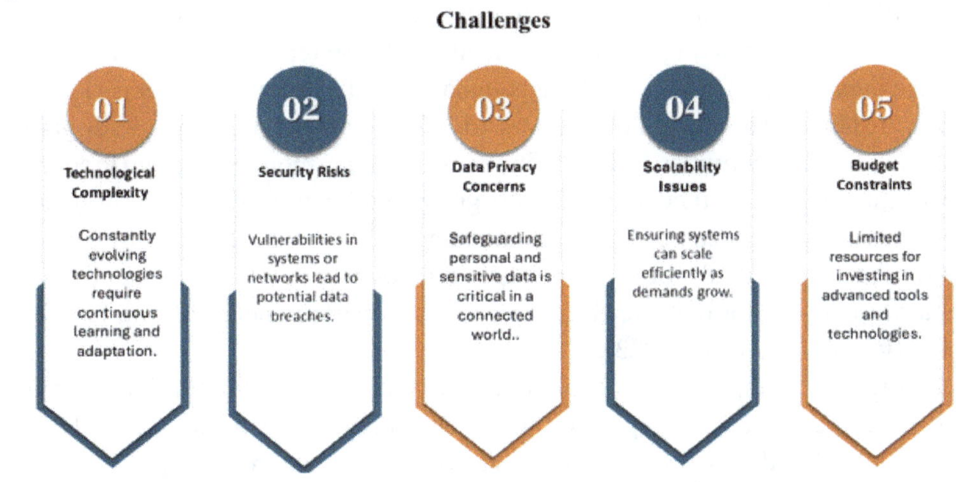

Ensuring inclusivity and accessibility in gamified experiences

When it comes to gamified experiences, making sure they work for everyone is crucial. I've personally found that some platforms just aren't friendly to all users, which makes you wonder—how do we make these things accessible? For one, games should always consider people with disabilities, like visual impairments or motor challenges. This can be done through options like screen readers, adaptable controls, or audio descriptions. But it's not just about disabilities; language and cultural differences can also be a hurdle. I remember trying to get into a game that was only available in one language—frustrating! Localization of content is one way to break those barriers. Then there's the issue of how games reward players. Some systems are overly complicated, leaving out those who might not have the time or skills to keep up. A balance between challenge and accessibility is essential. But let's not forget the importance of feedback from the players themselves. They're the ones experiencing it, and their insights are golden for improving inclusivity. So, making gamified systems inclusive isn't just about one thing—it's about considering all users, all the time.

7. FUTURE TRENDS IN GAMIFICATION

Emerging technologies (e.g., AR, VR) and their potential impact on gamification

Gamification is taking a thrilling turn, thanks to emerging technologies like augmented reality (AR) and virtual reality (VR) (Ravichandran et al., 2024; Brohi et al., 2020). Imagine walking through a historical site where digital characters pop up to share stories, making learning come alive. These technologies bring an immersive twist to gamification, pushing boundaries and enhancing user experiences.

For instance, AR apps allow players to blend real-world settings with game elements, making activities like education and training more engaging (Predescu & Mocanu, 2024; Chesti et al., 2020). VR goes even further, creating entirely new worlds where players can explore and interact, fostering deeper learning and retention (Lee et al., 2023). In corporate training, companies use VR simulations to prepare employees for real-life challenges in a safe environment, making learning more effective (Lee et al., 2023).

I remember trying out a VR training program at work, and it felt like stepping into a video game. It's exciting to see how such experiences can enhance skills and engagement. As AR and VR technologies advance, they will reshape gamification, making it more impactful and relevant (Familoni & Onyebuchi, 2024; Dogra et al.,

2021). The potential is immense, transforming how we learn and interact in various fields, from education to marketing and beyond. This blend of play and technology is where the future lies, making every interaction not just a task, but an adventure. Figure 7 shows consideration for gamification.

Figure 7. Consideration for Gamification

Predictions for the future of gamification in mobile apps

Gamification in mobile apps is shaping up to be an exciting trend for the future. As we look ahead, the role of gamification will likely grow in various areas. With more people using mobile devices, apps that make tasks fun will become more popular (Chan et al., 2024). Think about how a fitness app can motivate you with rewards for achieving your goals; that's gamification in action.

Personalization is also a key player in the future of gamification. Mobile apps will increasingly tailor experiences based on user preferences and behaviours, making users feel special and engaged (Khamaj & Ali, 2024). The blend of virtual reality (VR) and gamification can create immersive experiences that draw users in (Kong & Feng, 2024; Fatima-Tuz-Zahra et al., 2020).

Incorporating social elements will likely become more common, encouraging users to share achievements and compete with friends. This can create a sense of community and push users to engage more (Yassin, 2023).

Lastly, the focus on mental health and well-being through gamified experiences may gain traction. Apps that promote mindfulness or stress relief, using gamification to track progress, can have a positive impact on users' lives. With all these changes, it's clear that gamification will play a vital role in how we interact with mobile apps in the future.

The evolving landscape of user engagement strategies

Gamification is changing how companies engage users. Instead of just providing services, businesses are creating fun experiences to keep people coming back. For example, fitness apps often reward users with points or badges when they reach their goals. This makes exercising feel more like a game than a chore (Faizah et al., 2024).

Gamification is also being used in education. Teachers are incorporating elements like leaderboards and challenges to motivate students. When I taught a course, I noticed that students who competed for rewards were more involved and excited about learning. Using games in learning not only boosts participation but also helps students remember information better (Al-Abri et al., 2024).

Moreover, businesses are using gamification in marketing. Customers can earn points for making purchases or sharing content, turning shopping into a fun and rewarding activity (Faganel et al., 2024). This strategy not only increases sales but also builds a loyal customer base.

Lastly, technology is evolving, with new tools like virtual reality and augmented reality creating even more immersive gamified experiences. These trends show that gamification will likely continue to grow and adapt, engaging users in exciting ways that make them feel connected and invested in their experiences.

8. CONCLUSION

Summary of key points

This chapter explores how gamification enhances user engagement in learning apps, with techniques like points, badges, and leaderboards shown to increase motivation and encourage continued use. By tapping into basic psychological drivers—such as the need for achievement, connection, and personal growth—these game elements make learning feel more rewarding and enjoyable.

Real-world examples like Duolingo and Fitbit illustrate the power of gamification in creating engaging experiences that work across age groups and learning environments. These apps effectively use leaderboards and rewards to foster healthy competition and reinforce progress, keeping users actively involved in their learning journeys.

The chapter also addresses potential downsides, such as user burnout and diminishing returns when rewards lose their appeal over time. Practical advice on how to balance gamification with sustained engagement is offered to help developers design long-lasting, user-friendly experiences.

Finally, emerging technologies like AR and VR are discussed as exciting additions to gamification, suggesting that the future of digital learning may soon become even more immersive. This chapter inspires readers to consider innovative ways to make education both effective and enjoyable through thoughtfully crafted gamified experiences.

Final thoughts on the importance of gamification in enhancing user engagement

Gamification has proven to be more than just a trend; it's a powerful approach to keeping users engaged, motivated, and invested in their learning experiences. By using game-like elements, developers can transform what might otherwise feel like mundane tasks into interactive, rewarding activities. For many users, these small rewards—whether through points, badges, or the thrill of climbing a leaderboard—create a sense of accomplishment that drives them to keep going. Personally, I've seen how even subtle gamified elements make learning apps feel less like work and more like a challenge to be conquered.

The psychology behind gamification taps into universal motivations: we all want to feel a sense of progress and connection, especially in digital spaces. When apps incorporate rewards that align with users' goals and preferences, they create a sense of ownership and personal achievement. However, designing for sustained engagement is crucial. Balancing motivation with realistic pacing prevents burnout, ensuring that the experience remains enjoyable over time.

With technologies like AR and VR pushing boundaries, the potential for gamification to reshape digital learning is vast. In a world that increasingly values interactive and personalized experiences, gamification stands out as a vital tool, transforming how we learn, connect, and grow in the digital age.

REFERENCES

Al-Abri, A., Madiseh, F. R., & Moghaddam, M. M. (2024). Exploring Learning-Oriented Assessment in Enhancing Students' lexical Fluency through MALL. *The Asia-Pacific Education Researcher*. Advance online publication. DOI: 10.1007/s40299-024-00832-7

Alex, S. A., Jhanjhi, N., Humayun, M., Ibrahim, A. O., & Abulfaraj, A. W. (2022). Deep LSTM Model for Diabetes Prediction with Class Balancing by SMOTE. *Electronics (Basel), 11*(17), 2737. DOI: 10.3390/electronics11172737

Alferidah, D. K., & Jhanjhi, N. (2020). Cybersecurity Impact over Bigdata and IoT Growth. *2020 International Conference on Computational Intelligence (ICCI)*. DOI: 10.1109/ICCI51257.2020.9247722

Alkinani, M. H., Almazroi, A. A., Jhanjhi, N., & Khan, N. A. (2021). 5G and IoT Based Reporting and Accident Detection (RAD) System to Deliver First Aid Box Using Unmanned Aerial Vehicle. *Sensors (Basel), 21*(20), 6905. DOI: 10.3390/s21206905 PMID: 34696118

Ammattikorkeakoulu, J. (2015). *Gamification of the sales process at a telecommunications company to improve the motivation of the salesforce*. Theseus. https://www.theseus.fi/handle/10024/133103

Ananna, F. F., Nowreen, R., Jahwari, S. S. R. A., Costa, E. A., Angeline, L., & Sindiramutty, S. R. (2023). Analysing Influential factors in student academic achievement: Prediction modelling and insight. *International Journal of Emerging Multidisciplinaries Computer Science & Artificial Intelligence, 2*(1). Advance online publication. DOI: 10.54938/ijemdcsai.2023.02.1.254

Anvari, S. S., Hammer, J., & Wehbe, R. R. (2024). "More than just a game, it's an app that builds awareness around Mental Health": Mental Health Stigma Reduction Using Games for Change. *Proceedings of the ACM on Human-Computer Interaction, 8*(CHI PLAY), 1–30. DOI: 10.1145/3677090

Arroteia, N. F. P. (2024). Unveiling the invisible: Soft skills recognition from social media. *Company Games & Business Simulation Academic Journal, 4*(1), 69. DOI: 10.54988/cg.2024.1.1573

Audestad, J. A., & Telematikk, N. T. U. F. F. I. M. O. E. I. F. (2014). *The future of the mobile application market*. https://ntnuopen.ntnu.no/ntnu-xmlui/handle/11250/262967

Azam, H., Dulloo, M. I., Majeed, M. H., Wan, J. P. H., Xin, L. T., & Sindiramutty, S. R. (2023). Cybercrime Unmasked: Investigating cases and digital evidence. *International Journal of Emerging Multidisciplinaries Computer Science & Artificial Intelligence, 2*(1). Advance online publication. DOI: 10.54938/ijemdcsai.2023.02.1.255

Azam, H., Dulloo, M. I., Majeed, M. H., Wan, J. P. H., Xin, L. T., Tajwar, M. A., & Sindiramutty, S. R. (2023). Defending the digital Frontier: IDPS and the battle against Cyber threat. *International Journal of Emerging Multidisciplinaries Computer Science & Artificial Intelligence, 2*(1). Advance online publication. DOI: 10.54938/ijemdcsai.2023.02.1.253

Azam, H., Tajwar, M. A., Mayhialagan, S., Davis, A. J., Yik, C. J., Ali, D., & Sindiramutty, S. R. (2023). Innovations in Security: A study of cloud Computing and IoT. *International Journal of Emerging Multidisciplinaries Computer Science & Artificial Intelligence, 2*(1). Advance online publication. DOI: 10.54938/ijemdcsai.2023.02.1.252

Azam, H., Tan, M., Pin, L. T., Syahmi, M. A., Qian, A. L. W., Jingyan, H., Uddin, M. F., & Sindiramutty, S. R. (2023). Wireless Technology Security and Privacy: A Comprehensive Study. *Preprints.org*. DOI: 10.20944/preprints202311.0664.v1

Babbar, H., Rani, S., Masud, M., Verma, S., Anand, D., & Jhanjhi, N. (2021). Load balancing algorithm for migrating switches in software-defined vehicular networks. *Computers, Materials & Continua/Computers. Computers, Materials & Continua, 67*(1), 1301–1316. DOI: 10.32604/cmc.2021.014627

Bachoo, N., & Nagowah, L. (2024). GAMIFIE: A gamified framework as a strategy to increase motivation and engagement among employees in IT industries. In *Signals and communication technology* (pp. 95–118). DOI: 10.1007/978-3-031-57880-9_5

Baidya, R., Kumar, D., & Livvarcin, D. O. (2024). Gamified communication in social media for brand advocacy of Start-Ups. In *Routledge eBooks* (pp. 170–191). DOI: 10.4324/9781032694238-14

Bizzi, L. (2023). Why to gamify performance management? Consequences of user engagement in gamification. *Information & Management, 60*(3), 103762. DOI: 10.1016/j.im.2023.103762

Bolat, Y. I., & Taş, N. (2022). A meta-analysis on the effect of gamified-assessment tools' on academic achievement in formal educational settings. *Education and Information Technologies, 28*(5), 5011–5039. DOI: 10.1007/s10639-022-11411-y

Brohi, S. N., Jhanjhi, N., Brohi, N. N., & Brohi, M. N. (2020). Key Applications of State-of-the-Art Technologies to Mitigate and Eliminate COVID-19.pdf. *TECHRxiv*. DOI: 10.36227/techrxiv.12115596.v1

Brown, M., Bamberger, P., Bliese, P. D., & Shields, J. (2023). Fairness uncertainty and pay information exchange: Why and when employees disclose bonus pay to pay information websites. *Journal of Organizational Behavior*, *44*(9), 1362–1379. DOI: 10.1002/job.2739

Burghoorn, F., Scheres, A., Monterosso, J., Guo, M., Luo, S., Roelofs, K., & Figner, B. (2024). Pavlovian impatience: The anticipation of immediate rewards increases approach behaviour. *Cognitive, Affective & Behavioral Neuroscience*. Advance online publication. DOI: 10.3758/s13415-024-01236-2 PMID: 39467981

Cahya, A., & Indriana, D. (2023, August 30). *Transforming middle School mathematics education: The impact of gamification on student interest and engagement*. https://iocscience.org/ejournal/index.php/geneus/article/view/4749

Cao, Y., Gong, S., Wang, Y., Zheng, Q., & Wang, Z. (2022). How to provide competitors in educational gamification: The roles of competitor level and autonomous choice. *Computers in Human Behavior*, *138*, 107477. DOI: 10.1016/j.chb.2022.107477

Chan, G., Arya, A., Orji, R., Zhao, Z., & Whitehead, A. (2023). Increasing motivation in social exercise games: Personalising gamification elements to player type. *Behaviour & Information Technology*, •••, 1–31. DOI: 10.1080/0144929X.2023.2255293

Chan, G., Nwagu, C., Odenigbo, I., Alslaity, A., & Orji, R. (2024). The Shape of Mobile Health: A Systematic review of health visualization on mobile devices. *International Journal of Human-Computer Interaction*, •••, 1–19. DOI: 10.1080/10447318.2024.2313282

Chapman, J. R., Kohler, T. B., & Gedeborg, S. (2023). So, why do students perform better in gamified courses? Understanding motivational styles in educational gamification. *Journal of Educational Computing Research*, *61*(5), 927–950. DOI: 10.1177/07356331221127635

Chen, C., & Huang, K. (2024). The interplay of rewards and competition in digital game-based learning: Effects on intrinsic motivation, game performance and behaviors, and computational thinking. *Learning and Instruction*, *94*, 101995. DOI: 10.1016/j.learninstruc.2024.101995

Chesti, I. A., Humayun, M., Sama, N. U., & Jhanjhi, N. (2020). Evolution, Mitigation, and Prevention of Ransomware. *2020 2nd International Conference on Computer and Information Sciences (ICCIS)*. DOI: 10.1109/ICCIS49240.2020.9257708

Chiu, N. P. C., Rupitan, M. J., King, K. K., Sharbini, H., Borhan, N. H., & Yeo, A. (2024). Effects of Gamification in space Exploration learning. *Journal of Advanced Research in Applied Sciences and Engineering Technology*, 265–274. DOI: 10.37934/araset.59.1.265274

Christopoulos, A., & Mystakidis, S. (2023). Gamification in education. *Encyclopedia, 3*(4), 1223–1243. DOI: 10.3390/encyclopedia3040089

Cigdem, H., Ozturk, M., Karabacak, Y., Atik, N., Gürkan, S., & Aldemir, M. H. (2024). Unlocking student engagement and achievement: The impact of leaderboard gamification in online formative assessment for engineering education. *Education and Information Technologies, 29*(18), 24835–24860. Advance online publication. DOI: 10.1007/s10639-024-12845-2

Dastane, O., Fandos-Roig, J. C., & Sánchez-García, J. (2023). It's free! Still, would I learn? Unearthing perceived value of education apps for better entrepreneurial decisions. *Management Decision*. Advance online publication. DOI: 10.1108/MD-09-2022-1292

Demszky, D., Liu, J., Hill, H. C., Jurafsky, D., & Piech, C. (2023). Can automated feedback improve teachers' uptake of student ideas? Evidence from a randomized controlled trial in a Large-Scale online course. *Educational Evaluation and Policy Analysis, 46*(3), 483–505. DOI: 10.3102/01623737231169270

Denden, M., Tlili, A., Chen, N., Abed, M., Jemni, M., & Essalmi, F. (2022). The role of learners' characteristics in educational gamification systems: A systematic meta-review of the literature. *Interactive Learning Environments*, •••, 1–23. DOI: 10.1080/10494820.2022.2098777

Do, N., Jin, T., Priest, R., Meredith, L. N., & Landers, R. N. (2024). A longitudinal quasi-experiment of leaderboard effectiveness on learner behaviors and course performance. *Learning and Individual Differences, 116*, 102572. DOI: 10.1016/j.lindif.2024.102572

Dogra, V., Singh, A., Verma, S., Kavita, N., Jhanjhi, N. Z., & Talib, M. N. (2021). Analyzing DistilBERT for Sentiment Classification of Banking Financial News. In *Lecture notes in networks and systems* (pp. 501–510). DOI: 10.1007/978-981-16-3153-5_53

Ebadi, S., Rasouli, R., & Mohamadi, M. (2021). Exploring EFL learners' perspectives on using Kahoot as a game-based student response system. *Interactive Learning Environments, 31*(4), 2338–2350. DOI: 10.1080/10494820.2021.1881798

Eklund, A., Nygårds, A., & Smeds, D. (2018). *GeoFort: Mobile Game for Motivating Physical Activity using Gamification and Augmented Reality*. DIVA. https://www.diva-portal.org/smash/record.jsf?pid=diva2%3A1231096&dswid=-9383

Faganel, A., Pačarić, F., & Rižnar, I. (2024). The impact of gamification on Slovenian consumers' online shopping. *Administrative Sciences*, *14*(5), 86. DOI: 10.3390/admsci14050086

Faizah, A., Hardian, A. F. A., Nandini, R. D., Handayani, P. W., & Harahap, N. C. (2024). The influence of Incentive-Based Mobile Fitness apps on Users' continuance Intention with Gender Moderation Effects: Quantitative and Qualitative study. *JMIR Human Factors*, *11*, e50957. DOI: 10.2196/50957 PMID: 38837199

Familoni, N. B. T., & Onyebuchi, N. N. C. (2024). AUGMENTED AND VIRTUAL REALITY IN U.S. EDUCATION: A REVIEW: ANALYZING THE IMPACT, EFFECTIVENESS, AND FUTURE PROSPECTS OF AR/VR TOOLS IN ENHANCING LEARNING EXPERIENCES. *International Journal of Applied Research in Social Sciences*, *6*(4), 642–663. DOI: 10.51594/ijarss.v6i4.1043

Fatima-Tuz-Zahra, N., Jhanjhi, N., Brohi, S. N., Malik, N. A., & Humayun, M. (2020). Proposing a Hybrid RPL Protocol for Rank and Wormhole Attack Mitigation using Machine Learning. *2020 2nd International Conference on Computer and Information Sciences (ICCIS)*. DOI: 10.1109/ICCIS49240.2020.9257607

Gao, F., Yu, C., Zhou, H., & Xie, L. (2024). Behavioral lock screen application practice based on Gamification Design method. In *Lecture notes in computer science* (pp. 43–57). DOI: 10.1007/978-3-031-60898-8_3

Gil-Aciron, L. A. (2022). The gamer psychology: A psychological perspective on game design and gamification. *Interactive Learning Environments*, *32*(1), 183–207. DOI: 10.1080/10494820.2022.2082489

Gong, M., Wagner, C., & Ali, A. (2023). The impact of social network embeddedness on mobile massively multiplayer online games play. *Information Systems Journal*, *34*(2), 327–363. DOI: 10.1111/isj.12479

Gopi, R., Sathiyamoorthi, V., Selvakumar, S., Manikandan, R., Chatterjee, P., Jhanjhi, N. Z., & Luhach, A. K. (2021). Enhanced method of ANN based model for detection of DDoS attacks on multimedia internet of things. *Multimedia Tools and Applications*, *81*(19), 26739–26757. DOI: 10.1007/s11042-021-10640-6

Gouda, W., Almurafeh, M., Humayun, M., & Jhanjhi, N. Z. (2022). Detection of COVID-19 based on chest x-rays using deep learning. *Health Care*, *10*(2), 343. DOI: 10.3390/healthcare10020343 PMID: 35206957

Hall-Johnson, T., & Hall-Johnson, L. (2023). *Business model : FocusHome*. Theseus. https://www.theseus.fi/handle/10024/801120

Harbinger Group. (2024, April 18). *How to build a gamification Strategy – An expert view*. https://www.harbingergroup.com/blogs/how-to-build-a-gamification-strategy-an-expert-view/

Hellín, C. J., Calles-Esteban, F., Valledor, A., Gómez, J., Otón-Tortosa, S., & Tayebi, A. (2023). Enhancing Student Motivation and Engagement through a Gamified Learning Environment. *Sustainability (Basel)*, *15*(19), 14119. DOI: 10.3390/su151914119

Herbach, C. (2024). Placing people at the heart of customer experience journeys. In *Routledge eBooks* (pp. 65–92). DOI: 10.4324/9781003378099-5

Hong, Y., Saab, N., & Admiraal, W. (2024). Approaches and game elements used to tailor digital gamification for learning: A systematic literature review. *Computers & Education*, *212*, 105000. DOI: 10.1016/j.compedu.2024.105000

Huang, J., Chen, J., & Zhou, L. (2024). Motivation crowding effects on the intention for continued use of gamified fitness apps: A mixed-methods approach. *Frontiers in Psychology*, *14*, 1286463. Advance online publication. DOI: 10.3389/fpsyg.2023.1286463 PMID: 38268805

Humayun, M., Sujatha, R., Almuayqil, S. N., & Jhanjhi, N. Z. (2022). A Transfer Learning Approach with a Convolutional Neural Network for the Classification of Lung Carcinoma. *Health Care*, *10*(6), 1058. DOI: 10.3390/healthcare10061058 PMID: 35742109

Hunt, A., Merola, G. P., Carpenter, T., & Jaeggi, A. V. (2024). Evolutionary perspectives on Substance and Behavioural Addictions: Distinct and shared pathways to understanding, prediction and prevention. *Neuroscience and Biobehavioral Reviews*, *105603*, 105603. Advance online publication. DOI: 10.1016/j.neubiorev.2024.105603 PMID: 38402919

Hussain, K., Rahmatyar, A. R., Riskhan, B., Sheikh, M. U., & Sindiramutty, S. R. (2024). Threats and Vulnerabilities of Wireless Networks in the Internet of Things (IoT). *2024 IEEE 1st Karachi Section Humanitarian Technology Conference (KHI-HTC)*, *2*, 1–8. DOI: 10.1109/KHI-HTC60760.2024.10482197

Ibisu, A. E. (2024). Development of a gamification model for personalized e-learning. *arXiv (Cornell University)*. https://doi.org//arxiv.2404.15301DOI: 10.48550

Jansukpum, K., Chernbumroong, S., Intawong, K., Sureephong, P., & Puritat, K. (2024). Gamified Virtual Reality for Library services: The effect of gamification on enhancing knowledge retention and user engagement. *New Review of Academic Librarianship*, •••, 1–21. DOI: 10.1080/13614533.2024.2381509

Jha, A. K., & Verma, N. K. (2023). Social Media Platforms and User Engagement: A Multi-Platform Study on one-way firm sustainability communication. *Information Systems Frontiers*, *26*(1), 177–194. DOI: 10.1007/s10796-023-10376-8

Jhanjhi, N., Humayun, M., & Almuayqil, S. N. (2021). Cyber security and privacy issues in industrial internet of things. *Computer Systems Science and Engineering*, *37*(3), 361–380. DOI: 10.32604/csse.2021.015206

Juhász, L., Novack, T., Hochmair, H., & Qiao, S. (2020). Cartographic vandalism in the era of Location-Based Games—The case of OpenStreetMap and Pokémon GO. *ISPRS International Journal of Geo-Information*, *9*(4), 197. DOI: 10.3390/ijgi9040197

Jun, A. Y. M., Jinu, B. A., Seng, L. K., Maharaiq, M. H. F. B. Z., Khongsuwan, W., Junn, B. T. K., Hao, A. W., & Sindiramutty, S. R. (2024). Exploring the Impact of Crypto-Ransomware on Critical Industries: Case Studies and Solutions. *Preprints.org*. DOI: 10.20944/preprints202409.1325.v1

Karpouzis, K. (2024). What would Plato say? Concepts and notions from Greek philosophy applied to gamification mechanics for a meaningful and ethical gamification. *arXiv (Cornell University)*. /arxiv.2403.08041DOI: 10.31219/osf.io/u4cqn

Kastelli, S., & Takács, N. (2023). *Persuasive Language Learning - Qualitative research on user engagement in the persuasive system design of Duolingo*. DIVA. https://www.diva-portal.org/smash/record.jsf?pid=diva2:1773780

Khamaj, A., & Ali, A. M. (2024). Adapting user experience with reinforcement learning: Personalizing interfaces based on user behavior analysis in real-time. *Alexandria Engineering Journal*, *95*, 164–173. DOI: 10.1016/j.aej.2024.03.045

Khan, S. (2024). *Brave new words: How AI Will Revolutionize Education (and Why That's a Good Thing)*. Random House.

Kiyani, F. F., Hamid, B., Humayun, M., Sindiramutty, S. R. L., & Chowdhury, S. (2024). Discovery of Influential Publications Using Research Article's Usage Context. *2024 International Conference on Emerging Trends in Networks and Computer Communications (ETNCC)*, 1–7. DOI: 10.1109/ETNCC63262.2024.10767576

Kong, A., & Feng, Z. (2024). Advancing VR edutainment design in blended learning: Learners' views from wine classroom. *Computers & Education X Reality, 5*, 100078. DOI: 10.1016/j.cexr.2024.100078

Kumar, M. S., Vimal, S., Jhanjhi, N., Dhanabalan, S. S., & Alhumyani, H. A. (2021). Blockchain based peer to peer communication in autonomous drone operation. *Energy Reports*, *7*, 7925–7939. DOI: 10.1016/j.egyr.2021.08.073

Kunkel, T., Hayduk, T., & Lock, D. (2023). Push it real good: The effects of push notifications promoting motivational affordances on consumer behavior in a gamified mobile app. *European Journal of Marketing*, *57*(9), 2592–2618. DOI: 10.1108/EJM-06-2021-0388

Laine, T. H. (2018). Mobile Educational Augmented Reality Games: A systematic literature review and two case studies. *Computers*, *7*(1), 19. DOI: 10.3390/computers7010019

Landers, R. N., Bauer, K. N., Callan, R. C., & Armstrong, M. B. (2014). Psychological theory and the gamification of learning. In *Springer eBooks* (pp. 165–186). DOI: 10.1007/978-3-319-10208-5_9

Lee, S., Yang, Z., & Wu, J. G. (2023). Live, play, and learn: Language learner engagement in the immersive VR environment. *Education and Information Technologies*, *29*(9), 10529–10550. DOI: 10.1007/s10639-023-12215-4

Li, X., Yang, Y., & Chu, S. K. W. (2024). How does gamification bring long-term sustainable effects on children's learning? Implications from a crossover quasi-experimental study. *Educational Technology Research and Development*, *72*(3), 1357–1381. Advance online publication. DOI: 10.1007/s11423-023-10341-x

Lim, M., Abdullah, A., Jhanjhi, N., Khan, M. K., & Supramaniam, M. (2019). Link Prediction in Time-Evolving Criminal Network with deep Reinforcement learning technique. *IEEE Access : Practical Innovations, Open Solutions*, *7*, 184797–184807. DOI: 10.1109/ACCESS.2019.2958873

Linqiang, Y., Sindiramutty, S. R. L., Ashraf, H., Muzammal, S. M., Balakrishnan, S. P., Gupta, S., & Kavita, N. (2024). Intelligent Household Waste Classification System Based on Machine Learning. *2024 International Conference on Emerging Trends in Networks and Computer Communications (ETNCC)*, 760–768. DOI: 10.1109/ETNCC63262.2024.10767563

Mahmoudi-Dehaki, M., & Nasr-Esfahani, N. (2024). Harnessing the power of persuasion in education. In *Advances in educational technologies and instructional design book series* (pp. 177–204). DOI: 10.4018/979-8-3693-6397-3.ch008

Malik, G., Pradhan, D., & Rup, B. K. (2024). Gamification and customer brand engagement: A review and future research agendas. *Marketing Intelligence & Planning*. Advance online publication. DOI: 10.1108/MIP-12-2023-0659

Manchuri, A., Kakera, A., Saleh, A., & Raja, S. (2024). pplication of Supervised Machine Learning Models in Biodiesel Production Research - A Short Review. *Borneo Journal of Sciences and Technology*. DOI: 10.35370/bjost.2024.6.1-10

Mazlan, A. M. (2024). Savings vs Splurges : Mindful consumption for financial well-being. SSRN *Electronic Journal*. DOI: 10.2139/ssrn.4844943

McKay, E. N. (2013). *UI is Communication: How to design intuitive, user centered interfaces by focusing on effective communication*. https://dl.acm.org/citation.cfm?id=2513625

Mekler, E. D., Brühlmann, F., Opwis, K., & Tuch, A. N. (2013). *Do points, levels and leaderboards harm intrinsic motivation?* ACM., DOI: 10.1145/2583008.2583017

Mellado, R. (2024). *Principles of gamification for educational software*. Editora Dialética.

Modarelli, G., & Rainero, C. (2024). Perceptions, Criticalities, and Potentials of Game-Based Trainings for Inclusion. In *Advances in business information systems and analytics book series* (pp. 1–48). DOI: 10.4018/979-8-3693-1172-1.ch001

Mohammed, N., Saleem, A. N. S., & Anwar, M. (2024). Gamification application for promoting and encouraging physical activity in the elderly. eajse.tiu.edu.iq. DOI: 10.23918/eajse.v10i2p16

Moller, A. C., Kornfield, R., & Lu, A. S. (2024). Competition and Digital Game Design: A Self-Determination Theory Perspective. *Interacting with Computers*, iwae023. Advance online publication. DOI: 10.1093/iwc/iwae023

Morisio, M. (n.d.). *Design and implementation of a comfortable Mobile-based monitoring and recommendation system for healthy lifestyle habits - Webthesis*. https://webthesis.biblio.polito.it/31855/

Nayyar, A., Gadhavi, L., & Zaman, N. (2021). Machine learning in healthcare: review, opportunities and challenges. In *Elsevier eBooks* (pp. 23–45). DOI: 10.1016/B978-0-12-821229-5.00011-2

Neuschafer, T. (2023). *Understanding Duolingo Discussion Boards as Social-Emotional Support during the COVID-19 Pandemic*. https://eric.ed.gov/?id=EJ1393729

Nigam, A. C., & Chanda, R. S. (2024). Digital Innovations and their Effect on user Interaction in fitness Apps: A Comprehensive review and future research directions. *Advanced Series in Management*, 113–135. DOI: 10.1108/S1877-63612024000034B010

Nurcahyo, A. C. (2024). *Digital Literacy for Christian Students in Strengthening Faith within the Digital Realm*. AMERTA MEDIA.

Oke, A. E., Aliu, J., Mwanaumo, E. M. M., Abayomi, T., & Kahanji, C. (2023). Leveraging gamification to enhance productivity and employee engagement in the Nigerian construction industry. *Built Environment Project and Asset Management, 13*(6), 813–829. DOI: 10.1108/BEPAM-04-2023-0064

Ong, E. B. H., & Samson, B. P. V. (2024). Exploring design opportunities for improved self-motivation in self-tracking and health goal achievement. *Proceedings of the ACM on Human-Computer Interaction, 8*(MHCI), 1–17. DOI: 10.1145/3676501

Oravec, J. A. (2015). Gamification and multigamification in the workplace: Expanding the ludic dimensions of work and challenging the work/play dichotomy. *Cyberpsychology (Brno), 9*(3). Advance online publication. DOI: 10.5817/CP2015-3-6

Orji, R., Alslaity, A., & Chan, G. (2023). Towards understanding the mechanism through which reward and punishment motivate or demotivate behaviours. *Behaviour & Information Technology, 43*(6), 1042–1066. DOI: 10.1080/0144929X.2023.2196582

Pekpak, N., & Altunalan, T. (2024). Relationship Between Screen Time with Posture and Motor Coordination in Childhood: A Descriptive Study. *Turkiye Klinikleri Journal of Pediatrics, 33*(2), 41–47. DOI: 10.5336/pediatr.2024-101303

Petrou, S., Okonji, P., Olufemi, O., Kpokiri, E., Wood, C., Luedtke, S., Birjovanu, G., Aldosery, A., Ogunsola, F., & Kostkova, P. (2024). Gamified Antimicrobial Decision Support App (GADSA)-Is gamification more effective user engagement than group chat? *2024 IEEE 12th International Conference on Serious Games and Applications for Health (SeGAH)*, 1–7. DOI: 10.1109/SeGAH61285.2024.10639568

Pombo, N., & Santos, H. (2023). Lessons Learned from the Development of a Computerised Badge-based Reward Tool for Student Engagement in Learning Activities. *2023 IEEE World Engineering Education Conference (EDUNINE)*. DOI: 10.1109/EDUNINE57531.2023.10102849

Predescu, A., & Mocanu, M. (2024). Gamification in Real-World applications: interactive maps and augmented reality. In *IntechOpen eBooks*. DOI: 10.5772/intechopen.1004870

Ramadani, R. (2024). The impact of technology use on Young people: A case study of social media and internet usage. *Asian Journal of Research in Computer Science, 17*(8), 13–23. DOI: 10.9734/ajrcos/2024/v17i7486

Ravichandran, K., Virgin, B. A., Patil, S., Fatma, G., Rengarajan, M., & Bala, B. K. (2024). Gamifying Language Learning: Applying Augmented Reality and Gamification Strategies for Enhanced English Language Acquisition. *2024 Third International Conference on Smart Technologies and Systems for Next Generation Computing (ICSTSN)*, 1–6. DOI: 10.1109/ICSTSN61422.2024.10670787

Ravichandran, N., Tewaraja, T., Rajasegaran, V., Kumar, S. S., Gunasekar, S. K. L., & Sindiramutty, S. R. (2024). Comprehensive Review Analysis and Countermeasures for Cybersecurity Threats: DDoS, Ransomware, and Trojan Horse Attacks. *Preprints.org*. DOI: 10.20944/preprints202409.1369.v1

Rusin, N., & Szandała, T. (2024). The power of peer recognition points: Does it really boost employee engagement? *Strategic HR Review*. Advance online publication. DOI: 10.1108/SHR-06-2024-0040

Sanchez, E., Van Oostendorp, H., Fijnheer, J. D., & Lavoué, E. (2020). Gamification. In *Springer eBooks* (pp. 816–827). DOI: 10.1007/978-3-030-10576-1_38

Scanlon, D., Coulter, M., Baker, K., & Tannehill, D. (2024). The enactment of the Socially-Just Teaching Personal and Social Responsibility (SJ-TPSR) approach in physical education teacher education: Teacher educators' and pre-service teachers' perspectives. *Physical Education and Sport Pedagogy*, •••, 1–17. DOI: 10.1080/17408989.2024.2319071

Schluchter, T., Ahmed, N., & Huang, E. (2024). "It Actually Affected My Relationship": A Qualitative Analysis of Affordances and Attitudes Towards Mental Health Content on TikTok. *ACM, 5*, 491–507. DOI: 10.1145/3670653.3677501

Seng, Y. J., Cen, T. Y., Raslan, M. H. B. M., Subramaniam, M. R., Xin, L. Y., Kin, S. J., Long, M. S., & Sindiramutty, S. R. (2024). In-Depth Analysis and Countermeasures for Ransomware Attacks: Case Studies and Recommendations. *Preprints.org*. DOI: 10.20944/preprints202408.2261.v1

Shah, I. A., Jhanjhi, N. Z., & Laraib, A. (2022). Cybersecurity and blockchain usage in contemporary business. In *Advances in information security, privacy, and ethics book series* (pp. 49–64). DOI: 10.4018/978-1-6684-5284-4.ch003

Sharma, R., Singh, A., Kavita, N., Jhanjhi, N. Z., Masud, M., Jaha, E. S., & Verma, S. (2021). Plant disease diagnosis and image classification using deep learning. *Computers, Materials & Continua/Computers. Materials & Continua (Print)*, *71*(2), 2125–2140. DOI: 10.32604/cmc.2022.020017

Sharma, W., Lim, W. M., Kumar, S., Verma, A., & Kumra, R. (2023). Game on! A state-of-the-art overview of doing business with gamification. *Technological Forecasting and Social Change*, *198*, 122988. DOI: 10.1016/j.techfore.2023.122988

Sheng, L. T., Azman, A. B., Khan, N. A., Sindiramutty, S. R., Singh, S., & Verma, S. (2024). Reinforcement learning-based traffic signal control. *2024 International Conference on Emerging Trends in Networks and Computer Communications (ETNCC)*, 792–798. DOI: 10.1109/ETNCC63262.2024.10767501

Shortt, M., Tilak, S., Kuznetcova, I., Martens, B., & Akinkuolie, B. (2021). Gamification in mobile-assisted language learning: A systematic review of Duolingo literature from public release of 2012 to early 2020. *Computer Assisted Language Learning*, *36*(3), 517–554. DOI: 10.1080/09588221.2021.1933540

Sindiramutty, S. R., Jhanjhi, N., Tan, C. E., Lau, S. P., Muniandy, L., Gharib, A. H., Ashraf, H., & Murugesan, R. K. (2024). Industry 4.0. In *Advances in logistics, operations, and management science book series* (pp. 342–405). DOI: 10.4018/979-8-3693-1363-3.ch013

Sindiramutty, S. R., Jhanjhi, N. Z., Tan, C. E., Khan, N. A., Shah, B., & Manchuri, A. R. (2024). Cybersecurity measures for logistics industry. In *Advances in information security, privacy, and ethics book series* (pp. 1–58). DOI: 10.4018/979-8-3693-3816-2.ch001

Sindiramutty, S. R., Jhanjhi, N. Z., Tan, C. E., Khan, N. A., Shah, B., Yun, K. J., Ray, S. K., Jazri, H., & Hussain, M. (2024). Future trends and emerging threats in drone cybersecurity. In *Advances in information security, privacy, and ethics book series* (pp. 148–195). DOI: 10.4018/979-8-3693-0774-8.ch007

Sindiramutty, S. R., Jhanjhi, N. Z., Tan, C. E., Yun, K. J., Manchuri, A. R., Ashraf, H., Murugesan, R. K., Tee, W. J., & Hussain, M. (2024). Data security and privacy concerns in drone operations. In *Advances in information security, privacy, and ethics book series* (pp. 236–290). DOI: 10.4018/979-8-3693-0774-8.ch010

Sindiramutty, S. R., Tan, C. E., Shah, B., Khan, N. A., Gharib, A. H., Manchuri, A. R., Muniandy, L., Ray, S. K., & Jazri, H. (2024). Ethical considerations in drone cybersecurity. In *Advances in information security, privacy, and ethics book series* (pp. 42–87). DOI: 10.4018/979-8-3693-0774-8.ch003

Sindiramutty, S. R., Tan, C. E., & Wei, G. W. (2024). Eyes in the sky. In *Advances in information security, privacy, and ethics book series* (pp. 405–451). DOI: 10.4018/979-8-3693-0774-8.ch017

Singh, Y., & Milan, R. (2024). Utilitarian and hedonic values of gamification and their influence on brand engagement, loyalty, trust and WoM. *Entertainment Computing, 100868*. Advance online publication. DOI: 10.1016/j.entcom.2024.100868

Singhal, V., Jain, S. S., Anand, D., Singh, A., Verma, S., Kavita, N., Rodrigues, J. J. P. C., Jhanjhi, N. Z., Ghosh, U., Jo, O., & Iwendi, C. (2020). Artificial Intelligence Enabled Road Vehicle-Train Collision Risk Assessment Framework for Unmanned railway level crossings. *IEEE Access : Practical Innovations, Open Solutions, 8*, 113790–113806. DOI: 10.1109/ACCESS.2020.3002416

Siricharoen, W. V. (2023). Improving user experience (UX) by applying (Interactive) infographic in the human computer interaction context. *Mobile Networks and Applications*. Advance online publication. DOI: 10.1007/s11036-023-02179-7

Slamet, T., Brush, T., & Kwon, K. (2024). *The effects of competition in gamified online discussions on learners' behavioral and cognitive engagement*. Technology Knowledge and Learning., DOI: 10.1007/s10758-024-09740-7

Sotos-Martínez, V. J., Ferriz-Valero, A., García-Martínez, S., & Tortosa-Martínez, J. (2022). The effects of gamification on the motivation and basic psychological needs of secondary school physical education students. *Physical Education and Sport Pedagogy*, •••, 1–17. DOI: 10.1080/17408989.2022.2039611

Sukmana, Y., & Rosmansyah, Y. (2021). The use of Cloud FireStore for handling real-time data updates: An Empirical study of gamified Online quiz. *2021 2nd International Conference on Electronics, Communications and Information Technology (CECIT)*. DOI: 10.1109/CECIT53797.2021.00220

Temel, E. (2024, September 27). *Qualitative Analysis of Consumer's Responses to Mobile Push Notifications and Its Relationship with Impulsive Purchasing*. https://dergipark.org.tr/en/pub/jcsci/issue/87400/1557272

Tongpaeng, Y., Nobnop, R., Wongwan, N., Homla, P., Intawong, K., & Puritat, K. (2024). Comparison of gamified and non-gamified mixed reality in enhancing museum visitor engagement, motivation, and learning outcome. *Journal of Heritage Tourism, 19*(6), 1–30. DOI: 10.1080/1743873X.2024.2351852

Trinh, M. P., Chico, R. J., & Reed, R. M. (2023). How fun Overcame Fear: The gamification of a Graduate-Level Statistics course. *Journal of Management Education, 48*(4), 735–776. DOI: 10.1177/10525629231181120

Troussas, C., Krouska, A., & Virvou, M. (2017). Reinforcement theory combined with a badge system to foster student's performance in e-learning environments. *2017 8th International Conference on Information* [IISA]. *Intelligent Systems with Applications, 38*, 1–6. DOI: 10.1109/IISA.2017.8316421

Vaitone, N. V. (2025, May 27). -., Kirse, S., Sakalauske, K. a.-., Dikcius, V., & Zimaitis, I. (2024). The usefulness of gamification for enhancing customer loyalty to small e-tailers. *EuroMed Journal of Business, 20*(2), 435–452. Advance online publication. DOI: 10.1108/EMJB-09-2023-0240

Vashisht, D. (2023). Engaging and Entertaining customers: Gamification in interactive marketing. In *Springer eBooks* (pp. 807–835). DOI: 10.1007/978-3-031-14961-0_35

Waheed, A., Seegolam, B., Jowaheer, M. F., Sze, C. L. X., Hua, E. T. F., & Sindiramutty, S. R. (2024). Zero-Day Exploits in Cybersecurity: Case Studies and Countermeasure. *preprints.org*. DOI: 10.20944/preprints202407.2338.v1

Wang, W., & Sari, M. K. (2023). Examining the effect of the Task-Technology fit of game mechanisms on learning outcomes in online gamification platforms. *Journal of Educational Computing Research, 61*(8), 32–59. DOI: 10.1177/07356331231187285

Weiqi, X., Hooi, S. T. C., Sindiramutty, S. R. L., Asirvatham, D. L., Kumar, D., & Verma, S. (2024). Surface Anomaly Detection Using Machine Learning Technique. *2024 International Conference on Emerging Trends in Networks and Computer Communications (ETNCC)*, 1–7. DOI: 10.1109/ETNCC63262.2024.10767562

Wen, B. O. T., Syahriza, N., Xian, N. C. W., Wei, N. G., Shen, T. Z., Hin, Y. Z., Sindiramutty, S. R., & Nicole, T. Y. F. (2023). Detecting cyber threats with a Graph-Based NIDPS. In *Advances in logistics, operations, and management science book series* (pp. 36–74). DOI: 10.4018/978-1-6684-7625-3.ch002

Xi, N., & Hamari, J. (2019). Does gamification satisfy needs? A study on the relationship between gamification features and intrinsic need satisfaction. *International Journal of Information Management, 46*, 210–221. DOI: 10.1016/j.ijinfomgt.2018.12.002

Yassin, A. (2023). Exploring Gamification Advertising and its Role in Audience Engagement., *14*(1), 163–173. DOI: 10.21608/idj.2024.329267

Ying, X., Murugesan, R. K., Sindiramutty, S. R., Wei, G. W., Balakrishnan, S., Kumar, D., & Verma, S. (2024). Scene Text Recognition using Deep Learning Techniques. *2024 International Conference on Emerging Trends in Networks and Computer Communications (ETNCC)*, 1–9. DOI: 10.1109/ETNCC63262.2024.10767484

Zainuddin, Z., Chu, S. K. W., & Perera, C. J. (2024a). Gamification in Education. In *Sprinzer* (pp. 67–113). DOI: 10.1007/978-981-97-2219-8_3

Zainuddin, Z., Chu, S. K. W., & Perera, C. J. (2024b). Gamification Platforms for Flipped Learning Implementation. In *Springer* (pp. 167–180). DOI: 10.1007/978-981-97-2219-8_5

Zen, D. N., Wibowo, D. A., & Lesmana, I. (2024). The dangers of gadget addiction on the mental health of rural adolescents in Ciamis District. www.midwifery.iocspublisher.org. https://doi.org/DOI: 10.35335/midwifery.v12i4.1694

Zhang, J. (2023, June 1). *Gamification in marketing to increase customer retention*. https://dspace.mit.edu/handle/1721.1/151418

Zhang, J., Jiang, Q., Zhang, W., Kang, L., Lowry, P. B., & Zhang, X. (2023). Explaining the Outcomes of social Gamification: A longitudinal field experiment. *Journal of Management Information Systems*, *40*(2), 401–439. DOI: 10.1080/07421222.2023.2196776

Zhang, L., Shao, Z., Benitez, J., & Zhang, R. (2023). How to improve user engagement and retention in mobile payment: A gamification affordance perspective. *Decision Support Systems*, *168*, 113941. DOI: 10.1016/j.dss.2023.113941

Zosh, J. M., Lytle, S. R., Golinkoff, R. M., & Hirsh-Pasek, K. (2016). Putting the education back in Educational Apps: How content and context interact to promote learning. In *Springer eBooks* (pp. 259–282). https://doi.org/DOI: 10.1007/978-3-319-45102-2_17

Chapter 2

Optimizing Mobile Apps for Device-Free Indoor Localization in Dynamic Environments Using Analytical Geometry and Machine Learning

Ravikumar Ch
https://orcid.org/0000-0003-0809-5545
Sreenidhi University, Hyderabad, India

Vasepalli Kamakshamma
Srinivasa Ramanujan Institute of Technology, India

P. Radhika
TKR College of Engineering, India

Isha Batra
Lovely Professional University, India

Arun Malik
Lovely Professional University, India

Kalvog Prakasha Chary
CVR College of Engineering, India

ABSTRACT

Device-free indoor localization is pivotal in smart environments, enabling precise positioning for applications like smart homes, healthcare, and IoT systems. This chapter presents an advanced localization framework optimized for mobile apps, integrating analytical geometry with machine learning techniques to overcome the challenges of dynamic indoor environments. Using Received Signal Strength Indication (RSSI) data from strategically placed Wi-Fi modules, the system identifies precise locations without requiring individuals to carry devices. The methodology

DOI: 10.4018/979-8-3693-6925-8.ch002

includes systematic data collection, preprocessing, and training of Random Forest models, ensuring robustness against environmental changes such as moving obstacles and varying occupancy. Experimental results highlight significant improvements in accuracy and responsiveness compared to conventional approaches, establishing the proposed method as a viable solution for real-world indoor localization scenarios. This work offers a scalable and efficient pathway for mobile applications in dynamic IoT ecosystems.

1. INTRODUCTION

In recent years, the problem of indoor localization has garnered significant attention in research, primarily due to the growing use of indoor mobile robots for transportation and AI- driven communication services (Bécue, Praça, & Gama, 2021; Mirtaheri & Shahbazian, 2022). Unlike outdoor localization, where GPS-based methods are widely used, indoor environments present unique challenges that GPS cannot adequately address, such as the need for higher accuracy and the limited coverage range of indoor networks (Farahsari, Farahzadi, Rezazadeh, & Bagheri, 2022; Kordi, Alhammadi, Roslee, & Alias, 2020). Indoor localization technologies, including Wi-Fi, Bluetooth, Zigbee, Ultra Wideband (UWB), Radio Frequency Identification (RFID), Ultrasound, and

iBeacons, offer various solutions tailored to these challenges (Nouali, Moussaoui, & Derder, 2015; Mohanty, Tripathy, & Das, 2021; Tomic, Beko, & Dinis, 2022). Indoor positioning can be categorized into Line-of-Sight (LOS) and Non-Line-of-Sight (NLOS) scenarios, depending on the deployment and coverage range of access points (APs) (Li et al., 2020). The effectiveness of localization in these environments is influenced by factors such as obstacle number, thickness, and material, which contribute to signal attenuation and channel fading (Asaad & Maghdid, 2022; Barshandeh, Masdari, Dhiman, Hosseini, & Singh, 2021). Therefore, it is crucial to consider these attenuation factors when collecting distance indication data in different fading channel scenarios (Nouali, Moussaoui, & Derder, 2015; Fard Moshiri, Shahbazian, Nabati, & Ghorashi, 2021).

Received Signal Strength Indication (RSSI) is a key metric in indoor positioning, as it enables effective distance estimation from anchor points to a target location, particularly within theoretical distance-based models such as trilateration (Nouali, Moussaoui, & Derder, 2015; Mohanty, Tripathy, & Das, 2021). However, RSSI is often affected by various factors, including indoor interference, multipath effects, noise, and dynamic environmental conditions (Mahmood, Matin, Sarigiannidis, & Goudos, 2022). To enhance localization accuracy, recent studies have incorporated Channel State Information (CSI) alongside RSSI, as well as other parameters such as Round Trip Time (RTT), Direction of Arrival (DOA), and Time Difference of Arrival (TDOA) (Li et al., 2020; Asaad & Maghdid, 2022).

Indoor localization approaches can be broadly categorized into machine learning methods and filter-based techniques (Bellavista-Parent, Torres-Sospedra, & Pérez-Navarro, 2022). Traditional supervised machine learning methods, including Support Vector Machines (SVM), K-Nearest Neighbors (KNN), Naive Bayes, and decision trees, have been effective in data extraction, matching, and classification tasks related to localization (Mirtaheri & Shahbazian, 2022). As networks become increasingly complex, more advanced supervised learning techniques, such as Artificial Neural Networks (ANN), Convolutional Neural Networks (CNN), Deep Neural Networks (DNN), and Recurrent Neural Networks (RNN), have been developed (Nabati & Ghorashi, 2020). These methods focus on the training phase to map relationships between input and output layers through hidden layers, facilitating the creation of reliable models for online testing (Bellavista-Parent, Torres-Sospedra, & Pérez-Navarro, 2022).

Recent trends in machine learning-based localization have introduced the use of fingerprinting techniques, where collected RSSI data or virtual RSSI maps are matched against a pre-built fingerprint database (Fard Moshiri, Shahbazian, Nabati, & Ghorashi, 2021; Moradbeiki, Keshavarz, Rostami, Paiva, & Lopes, 2021). This approach often outperforms traditional trilateration, particularly in complex environments where signal interference is prevalent (Asaad & Maghdid, 2022). Unsupervised learning methods have also gained traction for their ability to dynamically update network parameters in real-time during the online phase, offering greater autonomy and reducing the need for extensive training (Bellavista-Parent, Torres-Sospedra, & Pérez-Navarro, 2022). These methods are particularly suited for scenarios requiring real-time adaptability without human intervention (Kordi, Alhammadi, Roslee, & Alias, 2020).

The major contributions of this paper are as follows:

a) The paper combines analytical geometry and machine learning, particularly Random Forest, to improve indoor localization accuracy without requiring devices on individuals.
b) It develops a scalable, low-power system using Wi-Fi modules for real-time location tracking in smart environments.
c) The system is proven to maintain high accuracy even with changing indoor conditions, making it suitable for practical IoT applications.

2. LITERATURE REVIEW

The landscape of indoor localization has evolved rapidly, driven by the increasing adoption of Internet of Things (IoT) technologies in smart city infrastructures and various other applications such as ambient assisted living, smart homes, and intelligent transportation systems. The growing complexity of these environments necessitates advanced localization techniques that can operate efficiently and accurately in real-time, adapting to the dynamic conditions typical of indoor settings. This literature review provides a comprehensive analysis of recent developments in indoor localization, particularly focusing on the integration of machine learning, hybrid approaches, and advanced mathematical models within IoT ecosystems.

2.1. The Role of Machine Learning in Indoor Localization

Machine learning (ML) has emerged as a pivotal technology in enhancing indoor localization systems, offering the capability to process and analyze large volumes of complex data. The integration of ML with IoT has enabled more intelligent and adaptive localization solutions, particularly in the context of smart cities. Wan et al. (2021) emphasized the role of ML in enabling intelligent vehicle location tracking in smart cities, where IoT data streams are processed to improve the precision of localization models (Wan, Zhang, Sun, & Wang, 2021). Their study illustrates how ML can be leveraged to handle the vast amount of data generated by IoT devices, ensuring that vehicle locations are accurately tracked even in the dense and complex urban environments characteristic of smart cities.

Dou et al. (2021) introduced a bisection reinforcement learning approach to 3D indoor localization, which significantly enhances the accuracy of position estimates in environments with complex indoor structures (Ravikumar, Kumar, Sathish, Suhasini, & Nimmala, 2024). This method demonstrates the potential of reinforcement learning (RL) to adaptively refine localization estimates based on real-time feedback, making it particularly suitable for applications requiring high precision in dynamically changing environments.

2.2. Hybrid and Multimodal Approaches

Hybrid localization approaches, which combine multiple localization techniques, have been developed to address the limitations of using a single method. These hybrid systems often integrate different types of sensors and data processing algorithms to improve accuracy and robustness in various scenarios. Ghorpade et al. (2021) proposed a hybrid optimized fuzzy threshold model combined with an Extreme Learning Machine (ELM) for the localization of elderly persons in IoT-based ambient assisted living environments (Ghorpade, Zennaro, & Chaudhari, 2021). This approach addresses the challenges of accurately tracking individuals in indoor settings, where factors such as signal attenuation and multipath effects can significantly impact the reliability of localization.

Thakur and Han (2021) explored multimodal approaches for indoor localization in smart homes, emphasizing the need for integrating multiple data sources to achieve reliable positioning (Jia, Zong, Huang, & Baker, 2021). Their work underscores the importance of combining data from various sensors, such as motion detectors, temperature sensors, and cameras, to create a more comprehensive and accurate localization system. By utilizing multiple data modalities, these systems can better cope with the inherent uncertainties and variabilities present in indoor environments, leading to more robust localization outcomes.

2.3. Fingerprinting Techniques and Machine Learning

Fingerprinting has become one of the most widely adopted techniques for indoor localization, particularly due to its effectiveness in environments where traditional methods like trilateration may struggle. Fingerprinting involves the creation of a database of signal characteristics (such as Received Signal Strength Indicator, or RSSI) at various locations within an environment. This database is then used to match real-time signal measurements to the pre-recorded fingerprints, thereby estimating the current location (Chen, Hsu, & Huang, 2021).

Jia et al. (2021) developed a Deep Neural Network (DNN)-based Wi-Fi RSSI fingerprinting method, which significantly enhances localization accuracy in IoT environments (Ravikumar, Naresh, Prasanna, Goud, & Prasad, 2024). Their approach leverages the power of deep learning to model the complex relationships between signal characteristics and spatial locations, allowing for more accurate position estimates even in the presence of noise and signal distortions. Similarly, Jain et al. (2021) explored the use of Bluetooth Low Energy (BLE) technology combined with RSSI fingerprinting and machine learning algorithms to achieve low-cost indoor localization in smart buildings (Kim, Nguyen, Kim, & Shim, 2021). This method provides a cost-effective solution for indoor localization, which is particularly appealing for large-scale deployments in smart buildings where budget constraints are a concern (Ashraf, Jhanjhi, Ihsan, Khan, & Humayun, 2025).

2.4. Advanced Mathematical Techniques: Low-Rank Matrix Completion and Differential Vectors

The application of advanced mathematical techniques has also played a crucial role in enhancing the performance of indoor localization systems. Low-rank matrix completion, for instance, is a technique used to handle incomplete data, which is a common challenge in IoT networks. Kim et al. (2021) proposed a deep learning-based low-rank matrix completion method that improves localization performance by accurately reconstructing missing data in IoT networks (Rana et al., 2024). This method is particularly useful in scenarios where sensor data may be lost or corrupted due to network issues or environmental interferences, as it allows for the recovery of missing information and thus maintains the integrity of the localization process.

Another innovative approach is the use of differential r-vectors for localization in IoT networks, as introduced by Tiwary et al. (2021) (Chen, Hsu, & Huang, 2021). This method enhances the robustness of localization systems by considering the differential characteristics of signal vectors, which can provide more reliable position estimates even in environments with significant signal fluctuations. By focusing on the differences between signal vectors rather than their absolute values, this approach reduces the impact of environmental noise and improves the accuracy of localization.

2.5. Deep Learning and Clustering Techniques in IoT Networks

Deep learning has proven to be highly effective in various aspects of indoor localization, particularly in clustering mobile sensor nodes within IoT networks. Spyridis et al. (2021) discussed the application of deep learning clustering techniques for tracing mobile sensor nodes in UAV networks, which are expected to play a critical role in the next generation of IoT systems (6G) (Ashraf, Jhanjhi, Ihsan, Khan, & Humayun, 2025). Their study highlights the potential of deep learning to manage the complex dynamics of mobile sensor networks, where the positions of nodes may change rapidly and unpredictably. By clustering sensor nodes based on their movement patterns and signal characteristics, deep learning algorithms can provide more accurate and efficient localization solutions in these highly dynamic environments.

2.6. Transfer Learning and Random Forest Approaches

Transfer learning, a machine learning technique where a model trained on one task is repurposed for another related task, has shown promise in the context of IoT-based localization. Chen et al. (2021) applied semi-supervised transfer learning for outdoor localization over LoRaWAN networks, demonstrating its effectiveness in reducing the need for extensive data collection when deploying localization systems in new environments (Zaheer, Tahir, Humayun, Almufareh, & Jhanjhi, 2022). This approach is particularly beneficial in scenarios where it is impractical or costly to gather large amounts of training data, as it allows for the reuse of existing models with minimal retraining.

Varma and Anand (2021) implemented a Random Forest-based learning approach for indoor localization in smart buildings, which effectively handles the high-dimensional data typically associated with IoT environments (Eldhai et al., 2025). Random Forest, an ensemble learning method, is well- suited for managing the complexity of IoT data, as it can handle a large number of input variables and interactions between them. This method has been shown to improve localization accuracy by making use of multiple decision trees to capture different aspects of the data, thereby providing a more comprehensive and reliable model for indoor positioning.

2.7. Emerging Challenges and Future Directions

Despite the significant advancements in indoor localization techniques, several challenges remain that need to be addressed to fully realize the potential of these systems in IoT environments. One of the primary challenges is the need for scalability, particularly as IoT networks continue to grow in size and complexity. As more devices and sensors are added to these networks, the demand for real-time processing and accurate localization increases, necessitating the development of more efficient algorithms and computational models.

Another challenge is the need for greater robustness and reliability, especially in environments where signal conditions may be highly variable. While techniques such as fingerprinting and machine learning have improved the accuracy of indoor localization, they are still susceptible to issues such as signal multipath effects, non-line-of-sight conditions, and environmental noise. Future research will need to focus on developing methods that can adapt to these challenges in real-time, ensuring consistent and reliable performance across a wide range of scenarios.

The integration of emerging technologies such as 6G, artificial intelligence, and advanced sensor networks will also play a crucial role in the future of indoor localization. As Spyridis et al. (2021) noted, the adoption of 6G technologies will enable new capabilities in IoT networks, such as the use of UAVs and other mobile platforms for dynamic localization (Shah, Jhanjhi, & Brohi, 2025). These advancements will require new approaches to data processing and sensor fusion, as well as the development of algorithms that can operate effectively in highly dynamic and complex environments.

Recent developments in machine learning have proved essential in tackling a variety of issues in the fields of technology, security, and health. In order to forecast health indicators, Ashraf et al. suggested a data-driven machine learning method that examines lifestyle factors like water intake, exercise, and sleep. As mentioned in (Varma & Anand, 2021), their work highlighted the potential of incorporating AI techniques to improve personalised health management. Similarly, in order to identify fake smart watch ads, Zaheer et al. presented a revolutionary machine learning methodology. This study demonstrated the value of strong AI models in guaranteeing the legitimacy of online marketplaces, offering a vital remedy for the expanding problem of dishonest digital advertising (Thakur & Han, 2021).

Additionally, studies have looked into creative approaches to improve software dependability and network security. In order to classify stream traffic in software-defined networks, Eldhai et al. used machine learning-based feature selection, emphasising how effective it is at enhancing data flow and system performance (Tiwary, Pandey, Kumar, & Youssef, 2021). By creating models for identifying software vulnerabilities using cutting-edge AI approaches, Shah et al. made a contribution to cybersecurity. Their results supported the creation of more secure web engineering frameworks by demonstrating the revolutionary role that generative AI plays in risk identification and mitigation (Ravikumar, Batra, & Malik, 2022). When taken as a whole, these examples demonstrate how AI may be successfully integrated into a variety of domains to effectively address urgent contemporary issues.

The growing need for precise and effective positioning solutions in smart cities and other applications is driving a dynamic and quickly changing sector, as evidenced by the literature on indoor localisation in IoT environments. The accuracy and dependability of localisation have significantly increased as a result of the use of machine learning, hybrid methodologies, and sophisticated mathematical techniques. The issues of scalability, resilience, and adaptability in these systems, however, require further study, especially as IoT networks develop and grow. The ongoing convergence of developing technologies is expected to influence indoor localisation in the future, providing fresh chances for creativity and improved performance in this crucial field of study (Ch, Batra, & Malik, 2021).

3. METHODOLOGY

The methodology for optimizing device-free localization in dynamic environments involves a systematic approach that integrates analytical geometry with machine learning techniques. The system is designed to accurately track individuals' locations within an indoor environment without requiring them to carry any devices. The localization process begins with the deployment of Wi-Fi modules, specifically

ESP8266 Wi-Fi modules, which are strategically placed throughout the monitoring area to form a wireless mesh network. These modules continuously monitor the RSSI (Received Signal Strength Indication) values, which fluctuate based on the presence and movement of individuals within the environment. These fluctuations in RSSI values are then used to estimate the location of the individuals.

To build a robust localization system, the methodology involves two key stages: data collection and data processing. During the data collection phase, the Wi-Fi modules are deployed in a controlled indoor environment, such as a laboratory. The environment is divided into a grid, with the modules placed at specific coordinates. As individuals move within this space, the RSSI data is recorded continuously, capturing the signal strength variations caused by their movement. This data collection process generates a dataset that is then divided into training and testing subsets. The training data is used to build and train machine learning models, while the testing data is used to evaluate the models' performance in accurately predicting locations.

Figure 1. System Workflow for Device-Free Localization.

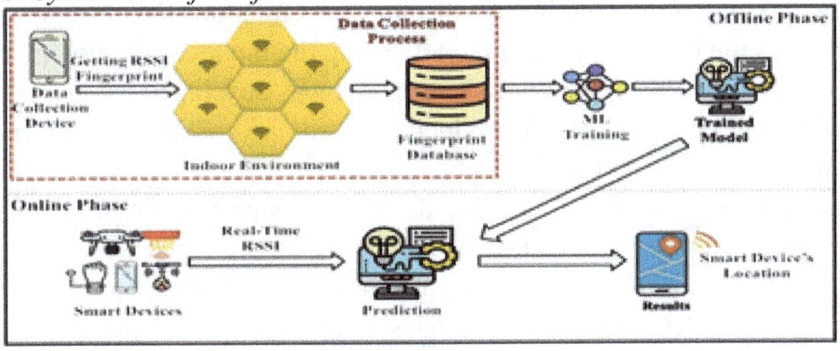

This figure illustrates the overall workflow of the device-free localization system, from data collection to real-time localization. The workflow highlights the interaction between the hardware components (ESP8266 Wi-Fi modules), the central processing unit, and the machine learning model used for location estimation.

Once the data is collected, it undergoes a series of processing steps to ensure it is ready for use in machine learning. This includes noise filtering, interpolation, and normalization. Noise filtering is applied to smooth out any short-term fluctuations in the RSSI values that may be caused by environmental noise. Interpolation is used to fill in any missing data points, ensuring a continuous data stream for analysis. Finally, normalization is performed to standardize the RSSI values, accounting for differences in signal strength across different sensors. This processed data is then organized into a matrix format, where rows represent the receiving nodes and columns represent the transmitting nodes. This matrix serves as the input for the machine learning models that are developed in the next phase.

4. IMPLEMENTATION

The implementation of the device-free localization system begins with the deployment of the ESP8266 Wi-Fi modules, which are configured to operate within the specified indoor environment. These modules are installed at predefined locations, forming a network that covers the entire monitoring area. Each module is configured to transmit and receive RSSI data continuously, sending this data to a central processing unit (CPU) for real-time analysis. The CPU is responsible for aggregating the RSSI data from all modules, processing it, and applying the machine learning model to estimate the individuals' locations within the environment.

To manage the flow of data and ensure efficient processing, a custom software application was developed. This application, running on the CPU, handles the collection, processing, and visualization of data. The Random Forest model, which was selected as the best-performing model during the model evaluation phase, was implemented within this application. The model was developed using Python and the TensorFlow framework, which provided the necessary tools for efficient machine learning model training and deployment.

A web-based user interface was also developed as part of the implementation. This interface allows users to monitor the localization system in real time, displaying the estimated locations of individuals as they move through the monitored area. The interface also provides performance metrics, such as accuracy and response time, giving users insights into how well the system is performing under different conditions. The integration of the Random Forest model with the real-time data from the Wi-Fi modules enables the system to provide accurate and timely location estimates, even in dynamic environments with multiple moving individuals.

5. RESULTS

The performance of the device-free localization system was evaluated through a series of experiments conducted in a controlled indoor environment. These experiments aimed to assess the system's accuracy, response time, and robustness to environmental changes. The Random Forest model, which was identified as the most effective during the model evaluation phase, demonstrated high accuracy in predicting the locations of individuals within the monitored area. The model outperformed other machine learning models, such as K-Nearest Neighbors (KNN) and Support Vector Machines (SVM), as shown in the performance metrics.

The accuracy of the localization system remained consistently high across various dynamic scenarios, including the presence of multiple individuals and changes in the environment, such as moving furniture or adding obstacles. The system's response time, which is the time taken to process the RSSI data and update the location estimate, was also measured. On average, the response time was approximately 2.5 seconds, which is sufficient for most real- time applications in indoor environments.

The system's robustness was further tested by introducing deliberate changes to the environment. Despite these changes, the Random Forest model maintained its accuracy, demonstrating the system's ability to adapt to varying conditions. The power consumption of the Wi-Fi modules was also monitored during the experiments. The modules consumed an average of 215 mA during signal transmission, making the system energy-efficient and suitable for continuous operation in real-world settings.

The overall results indicate that the proposed device-free localization system is effective in dynamic indoor environments. The integration of analytical geometry with machine learning, particularly the Random Forest model, significantly improves localization accuracy and robustness, making the system a viable solution for various IoT applications in smart environments.

Figure 2. Model Performance in Dynamic Environments.

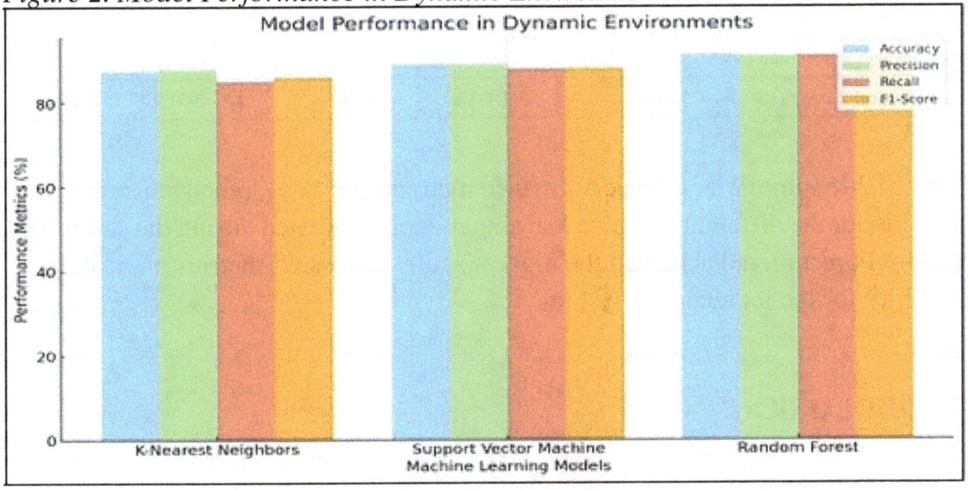

This figure compares the performance of different machine learning models (KNN, SVM, and Random Forest) in terms of localization accuracy across various dynamic scenarios. The Random Forest model shows superior performance, particularly in environments with significant changes.

Figure 3. Power Consumption of Wi-Fi Modules.

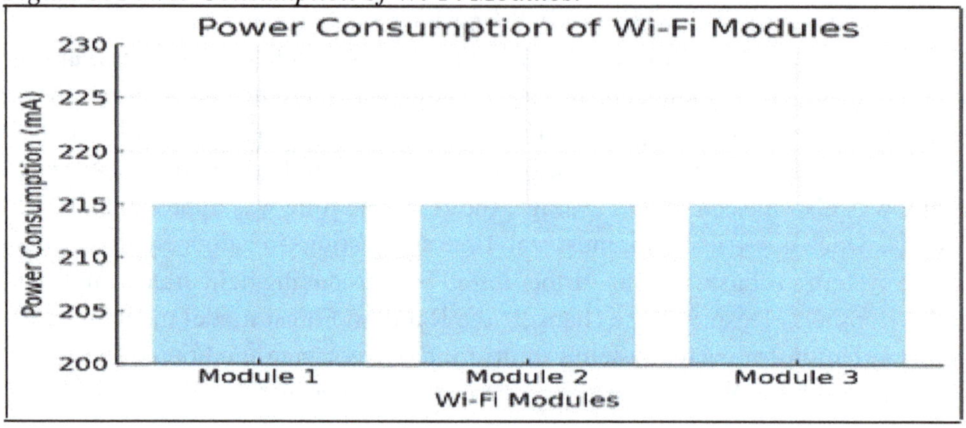

This figure illustrates the power consumption of the ESP8266 Wi-Fi modules during signal transmission. The figure shows that the modules consume an average of 215 mA, demonstrating the energy efficiency of the system.

Table 1. Model Performance Metrics

Model	Accuracy	Precision	Recall	F1-Score
K-Nearest Neighbors (KNN)	87.5%	0.88	0.85	0.86
Support Vector Machine (SVM)	89.0%	0.89	0.88	0.88
Random Forest (RF)	91.2%	0.91	0.91	0.91

This table summarizes the performance metrics (accuracy, precision, recall, and F1-score) for the different machine learning models evaluated during the study. The Random Forest model achieved the highest scores across all metrics, indicating its suitability for the localization system.

6. CONCLUSION

This study presents a comprehensive approach to optimizing device-free localization in dynamic indoor environments, using a combination of analytical geometry and machine learning techniques. The developed system, which relies on RSSI data collected from Wi-Fi modules, demonstrates a high level of accuracy in tracking individuals within various indoor settings. The Random Forest model, identified as the most effective among the tested algorithms, consistently outperformed others in terms of accuracy, precision, recall, and F1- score. Additionally, the system showed robustness to environmental changes, maintaining its performance even when the layout and occupancy of the environment were altered. The implementation details, including the workflow of data collection, processing, and model application, were carefully designed to ensure real-time responsiveness and energy efficiency, with power consumption metrics proving the system's suitability for continuous operation.

Overall, this research contributes a viable and scalable solution for device-free localization, with potential applications in smart homes, offices, and healthcare environments. Future work will focus on extending the system's capabilities by integrating additional sensor data and exploring the use of deep learning models to further enhance localization accuracy and adaptability.

REFERENCES

Asaad, S. M., & Maghdid, H. S. (2022). A comprehensive review of indoor/outdoor localization solutions in IoT era: Research challenges and future perspectives. *Computer Networks*, *212*, 109041. DOI: 10.1016/j.comnet.2022.109041

Ashraf, H., Jhanjhi, N. Z., Ihsan, U., Khan, N. A., & Humayun, M. (2025). Prediction of health indicators based on exercise, sleep, and water intake: Data-based machine learning approach. In *AI Techniques for Securing Medical and Business Practices* (pp. 291–310). IGI Global.

Barshandeh, S., Masdari, M., Dhiman, G., Hosseini, V., & Singh, K. K. (2021). A range-free localization algorithm for IoT networks. *International Journal of Intelligent Systems*, *37*(8), 10336–10379. DOI: 10.1002/int.22720

Bécue, I., Praça, I., & Gama, J. (2021). Artificial intelligence, cyber-threats and Industry 4.0: Challenges and opportunities. *Artificial Intelligence Review*, *54*(5), 3849–3886. DOI: 10.1007/s10462-020-09942-2

Bellavista-Parent, V., Torres-Sospedra, J., & Pérez-Navarro, A. (2022). Comprehensive analysis of applied machine learning in indoor positioning based on Wi-Fi: An extended systematic review. *Sensors (Basel)*, *22*(12), 4622. DOI: 10.3390/s22124622 PMID: 35746404

Ch, R., Batra, I., & Malik, A. (2021). Combining blockchain multi authority and botnet to create a hybrid adaptive crypto cloud framework. In *Proceedings of the 2021 International Conference on Computing Sciences (ICCS 2021)* (pp. 101–106). DOI: 10.1109/ICCS54944.2021.00028

Chen, Y. S., Hsu, C. S., & Huang, C. Y. (2021). A semi-supervised transfer learning with grid segmentation for outdoor localization over LoRaWANs. *Sensors (Basel)*, *21*(8), 2640. DOI: 10.3390/s21082640 PMID: 33918695

Dou, F., Lu, J., Xu, T., Huang, C. H., & Bi, J. (2021). A bisection reinforcement learning approach to 3-D indoor localization. *IEEE Internet of Things Journal*, *8*(8), 6519–6535. DOI: 10.1109/JIOT.2020.3041204

. Eldhai, A. M., Hamdan, M., Abdelaziz, A., Hashem, I., Marsono, M. N., Hamzah, M., & Jhanjhi, N. Z. (2025, February). Improved feature selection and stream traffic classification based on machine learning in software-defined networks. *IEEE Access*. IEEE.

Farahsari, P. S., Farahzadi, A., Rezazadeh, J., & Bagheri, A. (2022). A survey on indoor positioning systems for IoT-based applications. *IEEE Internet of Things Journal, 9*(10), 7680–7699. DOI: 10.1109/JIOT.2022.3149048

Fard Moshiri, P., Shahbazian, R., Nabati, M., & Ghorashi, S. A. (2021). A CSI-based human activity recognition using deep learning. *Sensors (Basel), 21*(21), 7225. DOI: 10.3390/s21217225 PMID: 34770532

Ghorpade, S. N., Zennaro, M., & Chaudhari, B. S. (2021). IoT-based hybrid optimized fuzzy threshold ELM model for localization of elderly persons. *Expert Systems with Applications, 184*, 115500. DOI: 10.1016/j.eswa.2021.115500

Jain, C., Sashank, G. V., Venkateswaran, N., & Markkandan, S. (2021). Low-cost BLE based indoor localization using RSSI fingerprinting and machine learning. In *Proceedings of the 2021 Sixth International Conference on Wireless Communications, Signal Processing and Networking (WiSPNET)* (pp. 363–367). Chennai, India. DOI: 10.1109/WiSPNET51692.2021.9419388

Jia, B., Zong, Z., Huang, B., & Baker, T. (2021). A DNN-based WiFi-RSSI indoor localization method in IoT. In *Springer Science and Business Media Deutschland GmbH* (Vol. 352, pp. 200–211). Berlin/Heidelberg, Germany. DOI: 10.1007/978-3-030-67720-6_14

Kim, S., Nguyen, L. T., Kim, J., & Shim, B. (2021). Deep learning based low-rank matrix completion for IoT network localization. *IEEE Wireless Communications Letters, 10*(10), 2115–2119. DOI: 10.1109/LWC.2021.3093983

Kordi, K. A., Alhammadi, A., Roslee, M., & Alias, M. Y. (2020). A review on wireless emerging IoT indoor localization. In *Proceedings of the 2020 IEEE 5th International Symposium on Telecommunication Technologies (ISTT)* (pp. 82–87). Shah Alam, Malaysia. DOI: 10.1109/ISTT50966.2020.9279386

Li, Y.. (2020). Toward location-enabled IoT (LE-IoT): IoT positioning techniques, error sources, and error mitigation. *IEEE Internet of Things Journal, 8*(5), 4035–4062. DOI: 10.1109/JIOT.2020.3021611

Mahmood, M. R., Matin, M. A., Sarigiannidis, P., & Goudos, S. K. (2022). A comprehensive review on artificial intelligence/machine learning algorithms for empowering the future IoT toward 6G era. *IEEE Access : Practical Innovations, Open Solutions, 10*, 87535–87562. DOI: 10.1109/ACCESS.2022.3199689

Mirtaheri, S. L., & Shahbazian, R. (2022). *Machine learning: Theory to applications.* CRC Press. DOI: 10.1201/9781003119258

Mohanty, S., Tripathy, A., & Das, B. (2021). An overview of a low energy UWB localization in IoT based system. In *Proceedings of the 2021 International Symposium of Asian Control Association on Intelligent Robotics and Industrial Automation (IRIA)* (pp. 293–296). Goa, India. DOI: 10.1109/IRIA53009.2021.9588718

Moradbeiki, A., Keshavarz, A., Rostami, H., Paiva, S., & Lopes, S. I. (2021). GNSS-free outdoor localization techniques for resource-constrained IoT architectures: A literature review. *Applied Sciences (Basel, Switzerland)*, *11*(23), 10793. DOI: 10.3390/app112210793

Nabati, M., & Ghorashi, S. A. (2020). Joint coordinate optimization in fingerprint-based indoor positioning. *IEEE Communications Letters*, *25*(5), 1192–1195. DOI: 10.1109/LCOMM.2020.3042375

Nouali, O., Moussaoui, S., & Derder, A. (2015). A BLE-based data collection system for IoT. In *Proceedings of the 2015 First International Conference on New Technologies of Information and Communication (NTIC)* (pp. 1–5). Mila, Algeria. https://doi.org/DOI: 10.1109/NTIC.2015.7368783

Rana, P., Batra, I., Malik, A., Ra, I.-H., Lee, O.-S., & Hosen, A. S. M. S. (2024). Efficacious novel intrusion detection system for cloud computing environment. *IEEE Access : Practical Innovations, Open Solutions*, *12*, 99223–99239. DOI: 10.1109/ACCESS.2024.3424528

Ravikumar, C., Batra, I., & Malik, A. (2022). A comparative analysis on blockchain technology considering security breaches. In *Lecture Notes in Networks and Systems* (Vol. 376, pp. 555–565).

Ravikumar, C., Kumar, K. H., Sathish, N., Suhasini, S., & Nimmala, S. (2024). A comprehensive analysis of machine learning methods for bug prediction in software development. *Lecture Notes in Electrical Engineering*, *1273*, 929–935. DOI: 10.1007/978-981-97-8031-0_99

Ravikumar, C., Naresh, B., Prasanna, P. L., Goud, E. A., & Prasad, P. R. (2024). Exploring machine learning algorithms for robust cyber threat detection and classification: A comprehensive evaluation. In *2024 International Conference on Intelligent Systems for Cybersecurity (ISCS).*

Shah, I. A., Jhanjhi, N. Z., & Brohi, S. N. (2025). Machine learning models for detecting software vulnerabilities. In *Generative AI for Web Engineering Models* (pp. 1–40). IGI Global.

Thakur, N., & Han, C. Y. (2021). Multimodal approaches for indoor localization for ambient assisted living in smart homes. *Information (Basel)*, *12*(3), 114. DOI: 10.3390/info12030114

Tiwary, P., Pandey, A., Kumar, S., & Youssef, M. (2021). Novel differential r-vectors for localization in IoT networks. *IEEE Sensors Letters*, *5*(6), 1–4. DOI: 10.1109/LSENS.2021.3083500

Tomic, S., Beko, M., & Dinis, R. (2022). *RSS-AoA-based target localization and tracking in wireless sensor networks*. CRC Press. DOI: 10.1201/9781003339335

Varma, P. S., & Anand, V. (2021). Random forest learning based indoor localization as an IoT service for smart buildings. *Wireless Personal Communications*, *117*(4), 3209–3227. DOI: 10.1007/s11277-020-07977-w

Wan, L., Zhang, M., Sun, L., & Wang, X. (2021). Machine learning empowered IoT for intelligent vehicle location in smart cities. *ACM Transactions on Internet Technology*, *21*(3), 1–25. DOI: 10.1145/3448612

Zaheer, A., Tahir, S., Humayun, M., Almufareh, M. F., & Jhanjhi, N. Z. (2022, November). A novel machine learning technique for fake smart watches advertisement detection. In *2022 14th International Conference on Mathematics, Actuarial Science, Computer Science and Statistics (MACS)* (pp. 1–5). IEEE. DOI: 10.1109/MACS56771.2022.10023151

Chapter 3
Assessing User Authentication and Authorization in Mobile Apps Ensuring Secure Access Control

Siva Raja Sindiramutty
https://orcid.org/0009-0006-0310-8721
Taylor's University, Malaysia

Noor Zaman Jhanjhi
https://orcid.org/0000-0001-8116-4733
Taylor's University, Malaysia

ABSTRACT

Mobile app security hinges on effective user authentication and authorization, yet we constantly see breaches due to weak safeguards. The chapter explores various methods to protect apps, from traditional password systems—often plagued by weak passwords and reuse—to advanced techniques like multi-factor and biometric authentication. But it's not just about getting people in; it's about making sure they access only what they should. Role-based and attribute-based access controls offer different ways to manage permissions, but if misconfigured, they can create loopholes. Throughout, smart tools play a big part in assessing these systems, catching flaws that developers may miss. Case studies drive home the risks, with real-world examples of fitness apps mishandling tokens or banking apps failing to secure biometric data. The chapter doesn't stop at the present; it also looks forward, examining how authentication and authorization will change with innovations like

DOI: 10.4018/979-8-3693-6925-8.ch003

password-less access and decentralized identities.

1. INTRODUCTION

Overview Of the Importance of Authentication and Authorization in Mobile Apps

Authentication and authorization play a huge role in keeping mobile apps secure. Without them, your data is out there for anyone to grab, especially considering how often we rely on apps for banking, communication, and even health monitoring (Nanda et al., 2023; Ananna et al., 2023). Imagine if someone had full access to your health app just because the security wasn't tight. I once downloaded an app that didn't require any serious login — I got a little paranoid and deleted it the same day. That's exactly why developers are always talking about how important it is to verify who's accessing the app, and then decide what they're allowed to do. Both authentication and authorization help to protect users from identity theft and other forms of attacks (Olabanji et al., 2024; Kiyani et al., 2024). Even small gaps in security can have devastating consequences. It's not just about keeping out hackers; it's about making sure legitimate users can trust the apps they use.

In recent times, apps have often targeted because they store so much sensitive data (Kanungo et al., 2024). A strong authentication process ensures that only verified users can log in, while proper authorization makes sure they only have access to what they're supposed to (Aboukadri et al., 2024; Kizza, 2024). This is especially important for apps handling financial transactions or personal health data, where the stakes are particularly high (Upadrista et al., 2023; Linqiang et al., 2024). We've all heard those stories about people losing everything to a simple app hack. No one wants to be that person. Honestly, I'm a lot more cautious now when I download apps.

A brief explanation of authentication (verifying user identity) vs. authorization (granting access to resources).

When I started using online services, I often got confused between authentication and authorization. It felt like they were the same thing, but they weren't. Authentication is all about confirming who you are. Think of it like entering your house — you use your key to prove it's your place. Whether you're logging into a social media account or accessing your work email, your password or fingerprint is your "key" (Alothman et al., 2023; Alferidah & Jhanjhi, 2020). On the other hand, authorization decides what you can do once you're inside. Imagine being at a party — you might be allowed in the house, but some rooms might be off-limits unless you're the host. This is how systems limit your access to only what you're permitted to see or do.

I remember a time at work when my login was successful, but I couldn't access a critical file. It wasn't a login problem — I just didn't have the right permissions, highlighting the difference between authentication and authorization. Even in cloud services, it's crucial to distinguish between these two terms. Misunderstanding this can lead to security issues, like letting unauthorized users access sensitive data. It might sound simple, but not knowing who can access what is one of the biggest security risks today. Figure 1 shows the importance of Authentication and Authorization in Mobile Apps.

Figure 1. Importance of Authentication and Authorization in Mobile Apps

Growing security challenges in mobile app ecosystems due to weak or outdated access control mechanisms.

A couple of years ago, I downloaded a fun mobile app that seemed harmless. But later, I found out it had some serious security flaws. This got me thinking about how risky mobile apps can be when access control isn't properly managed. Weak or outdated access control mechanisms are one of the growing concerns in mobile app ecosystems (Alothman et al., 2023; Azam, Dulloo, Majeed, Wan, Xin, & Sindiramutty, 2023). It's like having a lock on your front door but giving copies of the key to everyone in the neighbourhood. When apps fail to properly control who can access certain features or data, it opens huge security risks.

Developers sometimes rely on outdated methods or don't update their security protocols often enough, making it easier for attackers to exploit vulnerabilities. For instance, apps might store sensitive information in plain text or use simple authorization methods, which can be easily bypassed. I remember reading about a banking app that had outdated access controls — hackers were able to manipulate permissions, gaining unauthorized access to user data, which led to financial losses for many users.

In modern mobile environments, weak access control is a goldmine for attackers. As mobile devices are deeply integrated into our daily lives, attackers find more opportunities to exploit these weaknesses to steal personal and financial information (Hwaitat et al., 2024; Alkinani et al., 2021). Without regularly updated security measures, even trusted apps can become a potential threat to user privacy and safety (Zhou et al., 2018; Sheng et al., 2024).

Chapter Contribution

This chapter is pertinent to the area of cybersecurity in Assessing User Authentication and Authorization in Mobile Apps: Ensuring Secure Access Control. First, it describes Overview of the importance of authentication and authorization in mobile apps and outlines the Growing security challenges in mobile app ecosystems due to weak or outdated access control mechanisms. The chapter then investigates traditional password systems and their vulnerabilities, discusses the Pros and cons of enforcing password policies, How token-based systems reduce risks of session hijacking and credential exposure and Smart assessment tools for evaluating the accuracy and security of behavioural biometrics.

Furthermore, the chapter presents an Explanation of how RBAC and ABAC work, how Smart tools for assessing role assignments and potential risks of privilege escalation and Smart assessment methodologies for ensuring fine-grained control and preventing unauthorized access. It also discusses how assessment tools are used to test the robustness of mobile app authentication and authorization systems, and discusses on case study on Authentication and Authorization Failures in Mobile Apps. Lastly, the section provides a glimpse of How smart assessment tools will evolve to test these next-gen authentication methods and how smart assessment methodologies will adapt to decentralized authentication models.

Chapter Organization

in the first section, we will cover an insight into the introduction of authentication and authorization in mobile apps. In section 2 we go down common authentication methods in mobile apps. In section 3 we discuss the advanced authentication techniques followed by the authorization mechanisms in mobile apps section 4. Section 5 delves into the smart assessment tools for authentication and authorization. Section 6 will explain the case studies: authentication and authorization failures in mobile apps. In final section will discuss future trends in mobile app authentication and authorization.

2. COMMON AUTHENTICATION METHODS IN MOBILE APPS

Password-based authentication: Discussion on traditional password systems and their vulnerabilities (e.g., weak passwords, password reuse).

Password-based authentication is still widely used in mobile apps, even though it comes with several vulnerabilities (Kaur & Mustafa, 2019). Traditional password systems often rely on users creating passwords that are easy to remember, which sadly makes them easy to guess (Ur et al., 2016; Babbar et al., 2021). I remember one time, I reused the same password across multiple apps, and it only took one breach to expose my entire online identity. Weak passwords like "123456" or "password" are shockingly common and easily hacked, often leading to security breaches (Kammoun et al., 2022).

What's worse is the habit of reusing passwords. Users like me tend to repeat the same password across multiple platforms to avoid the hassle of remembering too many, but this creates a dangerous chain reaction—if one app gets hacked, all accounts using that password are at risk (Chakraborty et al., 2024; Azam, Dulloo, Majeed, Wan, Xin, Tajwar, et al., 2023). Password fatigue is a real issue, and when people reuse passwords or choose weak ones, it gives attackers an easier time breaking in (Prümmer, 2024; Brohi et al., 2020). Brute force attacks, where hackers try thousands of possible password combinations, also take advantage of predictable passwords (Raphael et al., 2024; Chesti et al., 2020).

Mobile apps that still depend on these systems expose users to unnecessary risks. I've often found myself clicking "Forgot Password" more than I should, which shows just how flawed the process can be. As much as passwords seem like an easy solution, they aren't enough to protect sensitive information, especially when users are left to create and remember them on their own (Kissell, 2024; Dogra et al., 2021).

Developers need to push for stronger alternatives or at least pair passwords with two-factor authentication (Blessing et al., 2024), making it harder for intruders to exploit these weaknesses. Figure 2 shows common authentication methods in mobile apps.

Pros and cons of enforcing password policies (complexity, length, expiration).

Enforcing password policies in mobile apps can be a double-edged sword. On one hand, strong policies requiring complex passwords, a mix of characters, numbers, and symbols can make accounts much harder to breach (He et al., 2024; Lykousas & Patsakis, 2024). I remember when one of my apps enforced a new password policy, and it made me feel more secure knowing my password wasn't just "abcd1234" anymore. Complexity rules help prevent common attacks like brute force, where hackers attempt countless combinations. But let's be real—complex passwords can also be a nightmare for users. People often struggle to remember these complicated combinations, leading them to either write passwords down or use "tricks" like adding a number at the end of a favourite password, which isn't much better (Kammoun et al., 2022b; Fatima-Tuz-Zahra et al., 2020).

Figure 2. Common Authentication Methods in Mobile Apps

Password length is another big factor. Longer passwords are statistically harder to crack, which sounds great in theory (Kävrestad et al., 2024; Azam, Tajwar, Mayhialagan, Davis, Yik, Ali, et al., 2023). But again, users often find longer passwords frustrating. I once set up a 16-character password for an app, and it was so painful to type it out every time, especially on a mobile keyboard. So, while longer passwords boost security, they can annoy users to the point where they use password managers—or worse, they just stop caring (Rooney et al., 2024; Gopi et al., 2021).

Then, there's the issue of password expiration. Frequent password changes are meant to improve security, but in practice, they often do more harm than good. People tend to create weaker passwords when they must reset them constantly (Paudel & Al-Ameen, 2024). I used to change mine every three months for work, and let's just say I got lazy, recycling the same pattern each time. The intention is good, but the execution often falls flat, making password policies a tricky balance between protection and usability.

Multi-Factor Authentication (MFA): Overview of MFA (e.g., SMS-based, app-based, and biometric factors).

When I first started using mobile apps, I thought my password was enough to keep things safe. But now, it feels like every app wants more security layers—and I get why. Mobile apps today often rely on **multi-factor authentication (MFA)** to secure user accounts (Wee et al., 2024; Gouda et al., 2022). It's become common practice for apps to ask for more than just a password. You've probably experienced MFA when you had to enter a code sent to your phone, whether it came through **SMS-based authentication**, or from a separate authentication app like Google Authenticator. While SMS feels convenient, it's not the safest, as hackers can intercept text messages (Lee, 2024; Azam, Tan, Pin, Syahmi, Qian, Jingyan, et al., 2023). Then there are **app-based methods**, which offer more protection since they generate codes directly on your phone(Wang et al., 2024). But, I'm not the best at keeping track of all these apps.

What blew my mind, though, was when my phone started asking for my **fingerprint**. Using **biometric authentication** is now super common in apps (Alrawili et al., 2024; Humayun et al., 2022). You just scan your face or fingerprint, and boom—you're in. It's fast and feels futuristic, even if sometimes my phone refuses to recognize my face when I first wake up. Still, it beats juggling multiple passwords. These layers make hacking harder for attackers, and as our reliance on mobile apps grows, it's comforting to know these safeguards are in place (Saadi et al., 2024). All these MFA methods aim to keep the user in control while making security more intuitive (Saadi et al., 2024b). I can't help but feel a little more secure swiping into my apps each day, even if I occasionally fumble with my fingerprint.

How smart assessment tools analyze the implementation of MFA for effectiveness.

In today's digital age, smart assessment tools are becoming essential for evaluating how effective multi-factor authentication (MFA) is in securing mobile apps (Mali, 2024; Hussain et al., 2024). These tools leverage advanced analytics to analyze MFA methods, ensuring they're not just there for show but providing the security needed. One approach is the **use of performance metrics**. Smart assessment tools track how users interact with MFA processes, noting how quickly they complete authentication steps (Karim et al., 2024; Jun et al., 2024). For instance, if users often drop out after entering their passwords, it's a sign that the process might be too cumbersome (Alajarmeh & Ladner, 2024; Jhanjhi et al., 2021). The data collected can show which MFA methods—like SMS codes, authenticator apps, or biometric options—lead to the quickest and most successful logins.

These tools can also evaluate **user feedback**. Gathering insights on user experiences can highlight which methods are user-friendly and which are causing frustration (Zhang et al., 2024; Manchuri et al., 2024). I once got locked out of an app because the biometric scanner wouldn't recognize my face after a long day, and I wasn't the only one; many users reported similar frustrations. Another fascinating aspect is the **behavioural analysis**. Smart tools can analyze patterns in how users authenticate themselves. For example, if someone typically logs in at a certain time and from a specific location but suddenly logs in from a different place, the system can flag this as a potential security risk, prompting additional authentication steps.

3. ADVANCED AUTHENTICATION TECHNIQUES

Token-based authentication: JSON Web Tokens (JWT) and OAuth2.

Token-based authentication, specifically using JSON Web Tokens (JWT) and OAuth2, has become popular for securing web applications (Prinakaa et al., 2024; Kumar et al., 2021). JWT is like a tiny, digital ID card, containing encoded data that lets servers know who you are without needing to constantly check a database. It's a simple concept—when I first started learning about it, I was surprised at how efficient it makes things. You send the token, the server trusts you, and voila! You're in. But of course, not all that glitters is gold. JWT tokens, while convenient, can be risky if not secured properly—token leakage could lead to unauthorized access (Bucko et al., 2023; Ravichandran et al., 2024).

OAuth2, on the other hand, is more like a valet key. It lets you access certain parts of a system without giving away full control. Think about how Google or Facebook lets you sign in to a third-party app without needing to create a new password. It's magic until you realize that, once you grant permission, you're handing over trust to those platforms, too. The blend of OAuth2 and JWT works like a charm in modern apps—when done right. The real trick is in balancing usability and security because, while OAuth2 simplifies access, developers must stay sharp to avoid potential pitfalls.

Interestingly, some apps combine both methods to cover all angles. I've seen this in some of my projects—OAuth2 handles the external logins, while JWT manages internal sessions. It makes you appreciate how much we rely on these "keys" to keep things moving smoothly (Anderson, 2023; Seng et al., 2024).

How token-based systems reduce risks of session hijacking and credential exposure.

Token-based authentication is a great way to improve security, especially when it comes to preventing session hijacking and protecting credentials (Alassery et al., 2015; Sindiramutty, 2024)

Here's how it works: when you log in, instead of constantly sending your password over the network, you receive a token—a unique string of numbers or characters. This token acts as a key to unlock various resources during your session (Khamisa, 2021; Lim et al., 2019). If someone intercepts that token, the damage is limited because tokens expire after a short period or can be invalidated when suspicious activity is detected.

Now, think about how terrifying session hijacking can be. Imagine someone secretly taking control of your account while you're still logged in—hijacking your session, so to speak. Token-based systems significantly lower this risk by using methods like encryption and limited token lifespans (Yang et al., 2024; Sindiramutty, Prabagaran, Akbar, Hussain, & Malik, 2024). When tokens are securely stored and regularly refreshed, attackers have a much smaller window to exploit them (Aghili, 2024).

Credential exposure is another nightmare. Passwords, especially when reused across platforms, are vulnerable (Oliveira et al., 2024; Sindiramutty, Prabagaran, Jhanjhi, Ghazanfar, Malik, & Soomro, 2024). With token-based authentication, your password isn't repeatedly sent with every request—just once during login (Jiang et al., 2023; Nayyar et al., 2021). This reduces the chances of your credentials being exposed if the connection is compromised. I remember once leaving a laptop at a café, and logging into a site. Realizing how token expiration saved me from a potential breach still gives me chills.

By keeping tokens short-lived and invalidating them quickly in suspicious cases, token-based systems strike a balance between user convenience and security. It's a simple but effective solution to a complex problem. Figure 3 shows types of advanced authentication techniques.

Figure 3. Types of Advanced Authentication Techniques

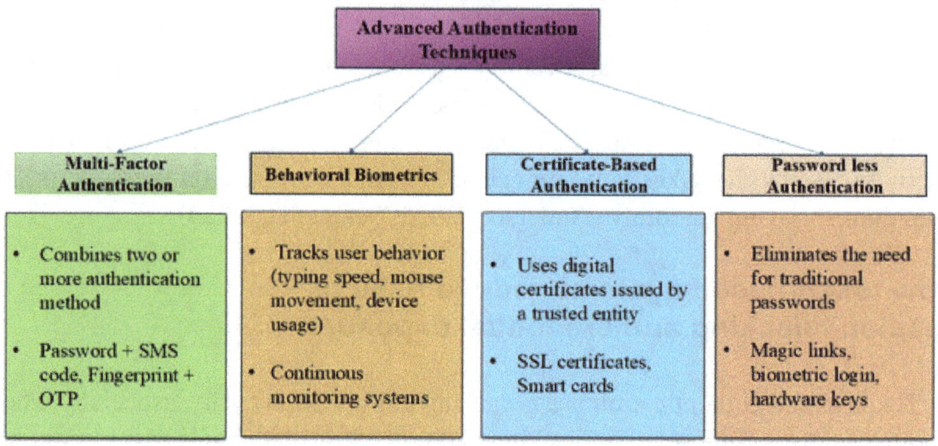

Assessment of token lifetimes, revocation, and storage mechanisms.

Token lifetimes, revocation, and storage mechanisms are crucial in maintaining a secure token-based authentication system (Rabiah et al., 2024). The balance between usability and security often lies in how long tokens remain valid and how efficiently they can be revoked (Jamwal et al., 2024). A token's lifetime defines how long it stays valid before requiring renewal or reauthentication (Gnanasekaran, 2024; Shah et al., 2022). Shorter lifetimes add a layer of security, as they limit the window of opportunity for an attacker to misuse a stolen token. But it also means more frequent logins, which can frustrate users. I remember using a platform that asked me to log in almost every hour—it was safe but exhausting.

Revocation is another critical element. If a token is compromised, the system needs a way to invalidate it instantly (Noble & Corner, 2002; Sharma et al., 2021). Systems often use a blacklist or token revocation list to ensure that stolen or expired tokens can no longer grant access (Kim et al., 2024). Once, I had an app revoke a token while I was on vacation. At first, it seemed inconvenient, but then I realized how much worse it could have been if someone had accessed my account.

Storage mechanisms for tokens also play a big part. If tokens aren't stored securely, they become easy targets for attackers (Trampert et al., 2023). Most systems use secure storage methods, such as encryption or hardware-backed key storage, to keep tokens safe (Bag et al., 2023). Local storage on the client side can be risky, especially if tokens are stored in plaintext or poorly secured environments. On the server side, it's important to keep access logs and security measures tight to prevent unauthorized access.

Managing token lifetimes, efficient revocation, and secure storage make a world of difference in protecting sensitive data.

Behavioural Biometrics

Using patterns like typing speed, touch gestures, and movement as additional authentication factors.

When I first heard about behavioural biometrics, I was amazed at how much it relies on patterns in our everyday actions. Imagine, something as ordinary as the way you type or swipe your phone could help prove who you are. Typing speed, for example, varies between people. It's almost like a digital fingerprint, as we all have a natural rhythm when we press keys. Researchers have found that analysing these patterns adds an extra layer of security (Gupta et al., 2023; Salem et al., 2022). It feels more personal than just typing in a password or swiping a fingerprint.

Touch gestures are another interesting angle. Think about how you naturally scroll through a webpage or swipe to unlock your phone. It turns out we each have slight variations in how we move our fingers, and companies are now tapping into these habits to build stronger authentication systems (Gupta et al., 2023b; Singhal et al., 2020). Even how you hold your phone, or walk can be a giveaway of your identity. Your device can sense those subtle movements, and it's becoming another checkpoint for verifying that you're you.

What's fascinating is the potential these techniques must reduce fraud and secure sensitive information. Especially in banking, where multiple layers of security are crucial, behavioural biometrics can offer that extra safeguard. It's like adding locks to a door, only they're invisible—just embedded in your everyday actions.

Smart assessment tools for evaluating the accuracy and security of behavioral biometrics

Evaluating behavioural biometrics with smart assessment tools feels like managing a constantly moving target. These biometrics, such as keystroke patterns, mouse movements, or gait, rely on how we behave—uniquely different from one person to the next (Khan et al., 2024). But how do you measure the accuracy and security of something fluid, changing subtly over time? Well, smart assessment tools attempt to do just that. They aim to track tiny shifts in user behaviour without becoming too forgiving or overly strict. I remember reading about how systems sometimes misinterpret legitimate users if they type a little slower due to fatigue or stress.

These tools often incorporate machine learning algorithms, which help recognize behavioural traits and catch potential fraud (Josyula, 2023). But the tech isn't perfect. There's always a balance between tightening security and keeping the user experience smooth (Asatiani et al., 2023). I've seen cases where people got locked out of their accounts just because they were typing differently after a long day. While advancements continue, there's no single perfect solution. Instead, these tools combine multiple layers of checks, ensuring behaviour is continuously monitored in a way that's hard to fake (Bhatt et al., 2023). As smart as they are, they still need a human touch to address unexpected quirks and the unpredictability of human behaviour (Bloom, 2023). So, the future looks promising but not without its bumps along the way

4. AUTHORIZATION MECHANISMS IN MOBILE APPS

Role-based Access Control (RBAC):

Explanation of how RBAC works (assigning permissions based on user roles).

Role-based access control (RBAC) is an essential part of keeping mobile apps secure (Saxena & Alam, 2023; Sindiramutty, Prabagaran, Jhanjhi, Murugesan, Brohi, & Masud, 2024). I've noticed, especially when managing my app accounts, that certain features are blocked depending on my role in the system—like how I couldn't access admin settings when using my work profile on our company app. That's RBAC in action. It assigns permissions based on user roles instead of granting blanket access to everyone (Kizza, 2024). So, if you're a regular user, you only get access to the necessary functions. This method limits potential damage if a user's credentials get compromised. It's the idea of giving just enough access to get the job done, and nothing more.

In mobile apps, this control means assigning permissions that are appropriate for the role—whether that's admin, editor, or viewer. For instance, think about banking apps: only the customer support agent can make account changes, while regular users can only check balances and transfer funds. RBAC offers a more straightforward, less risky way to manage access across a variety of users without needing to manually adjust individual permissions (Yang, Zhao, et al., 2024). Plus, it's easier for developers too. You set up roles once and then assign them instead of constantly tweaking individual permissions. I remember working on a project where a similar system saved us a ton of headaches, especially when onboarding new users. Figure 4 shows the types of authorization mechanisms.

Figure 4. Types Of Authorization Mechanisms

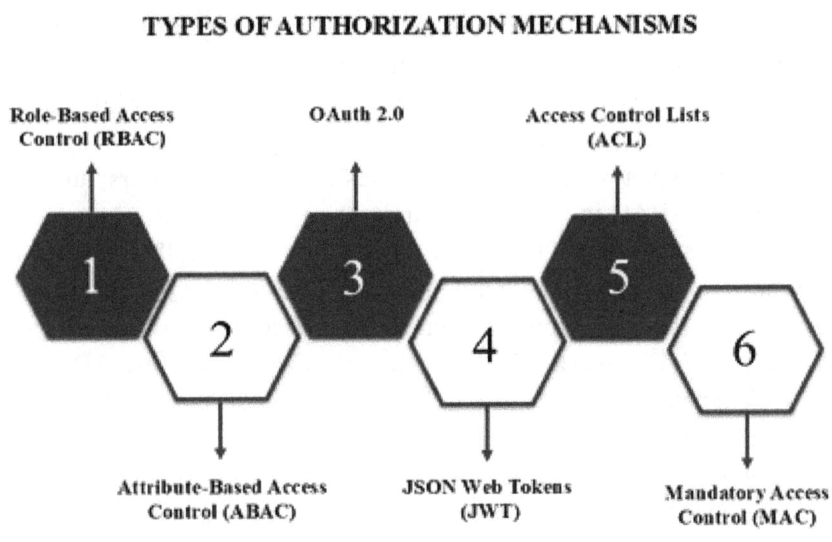

Smart tools for assessing role assignments and potential risks of privilege escalation.

Smart tools for assessing role assignments and the risks of privilege escalation are becoming essential, especially as mobile apps and systems grow more complex. These tools help track who has access to what and ensure that roles are assigned properly, without leaving room for unauthorized actions. A key issue is privilege escalation, where someone gains more access than they should. Think of it like a guest at a party suddenly finding themselves in the VIP area without an invite. It's a problem that can lead to major security breaches. Advanced tools even use AI and machine learning to predict possible future risks, analyzing user behaviour patterns to identify unusual access requests (Alzaabi & Mehmood, 2024). This proactive approach can help prevent privilege escalation by flagging potential threats early. It's like having a security guard who not only watches the doors but also predicts when someone might try to sneak in through the window. With regular monitoring, smart tools help apps maintain tight security, making it harder for privilege escalation to occur.

Attribute-based Access Control (ABAC): Introduction to ABAC (granting access based on user attributes like location, device, or time).

Attribute-Based Access Control (ABAC) is like giving someone a key based on who they are, where they are, and what time it is (Alzaabi & Mehmood, 2024). Instead of the usual "yes" or "no" access based on roles, ABAC digs a bit deeper into specific details. Imagine this: I once tried logging into my work system from home, but my access was denied because it was past working hours, and I was on my laptop. This is ABAC in action—ensuring the right conditions are met before granting access (Sarfaraz et al., 2023; Sindiramutty, Prabagaran, Jhanjhi, Murugesan, Brohi, & Wei, 2024). It considers factors like location, device type, and time of day. This makes ABAC flexible, allowing organizations to control access more precisely (Choksy et al., 2023; Pham et al., 2023). Hospitals, for example, use ABAC to limit sensitive medical data access only to staff in specific roles, on secure devices, during working hours (J. A. Khan, 2024)

In the real world, ABAC reduces risks by blocking unauthorized users even when they know passwords (Alsmadi, 2023). Imagine the damage if a hacker got into the system during non-business hours. ABAC's strength lies in its ability to tighten security without adding too much complexity for users. As technology keeps getting more sophisticated, the way we protect sensitive data must also keep up. And with ABAC, access is not just about who you are, but also about where, when, and how you're accessing the system (Najat et al., 2023).

Smart assessment methodologies for ensuring fine-grained control and preventing unauthorized access.

Smart assessment methodologies aim to give organizations a way to carefully monitor and manage who has access to their data, going beyond the basic yes-or-no checks. Imagine a smart office where your badge doesn't just let you in the building but also limits where you can go based on your role, the time, or even if you're using a company device. That's how smart assessments work in the digital world. They look at multiple factors—user roles, behaviours, and even the device being used—to ensure fine-grained control over who gets to access what.

I once saw a colleague's access to a critical system blocked just because they logged in from an unrecognized device. This felt like an extra hassle at the time, but it's the kind of smart assessment that can prevent data breaches. These methodologies are designed to analyze user attributes in real time, meaning that they continuously assess whether a user should maintain access as their situation changes.

5. SMART ASSESSMENT TOOLS FOR AUTHENTICATION AND AUTHORIZATION

Overview of smart assessment tools used to test the robustness of mobile app authentication and authorization systems.

Testing the mobile app authentication and authorization system, on the other hand, has been found to need smart assessment tools as well (Saqib & Moon, 2022; Sindiramutty, Prabagaran, Jhanjhi, Murugesan, Malik, & Hussain, 2024). "Getting hacked is like leaving your front door open!" A friend once told me. He wasn't wrong. Tools on the other hand are used to confirm that everything has been properly secured so vulnerabilities do not provide an open door for hackers. A widely used tool is OWASP ZAP, designed to detect vulnerabilities in mobile authentication and verify users (Domoney 2024). Here you have it, Burp Suite found some security issues with the authorization and restricted users from their expected access to only those they are supposed to wear (Wear 2018).

A more recent tool which is gaining attention nowadays is MobSF (Zirillo et al., 2024). In (Archibong et al., 2024), MARA is introduced which audits for improper configurations and vulnerabilities in mobile apps. Kind of like locking the door but forgetting to make sure the windows are closed — MobSF does that extra check. The problem with this is that some of these tools surprisingly quickly identify errors which may seem minor but cause massive issues. Recently, the Firebase Scanner available on GitHub was introduced to detect authentication vulnerabilities (Devi et al., 2023). This isn't just about developers being protected; it's about the end user's data potentially getting stolen. Figure 5 Want to know more about small assessment tools?

Figure 5. Key Features of Small Assessment Tools

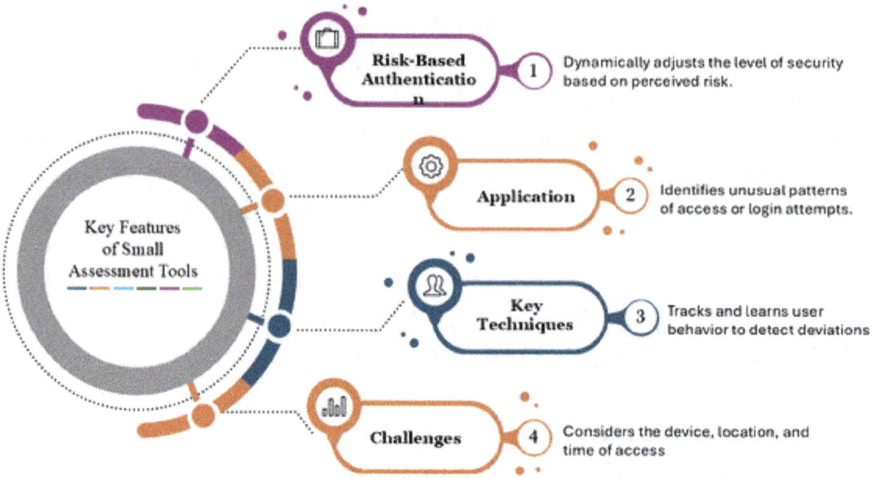

Penetration testing tools

Automated tools for brute-force password attacks and token manipulation.

To guarantee secure authentication and authorization processes, penetration testing tools are vital (Igwenagu et al., 2024). These tools typically provide automated brute-force password attacks and token manipulation solutions to discover weaknesses before they are abused by attackers. For example, brute-force attacks are automatically performed using tools such as Hydra and Burp Suite to try out lots of password combinations in rapid succession against different services (Kaushik et al., 2024; Waheed et al., 2024). This approach saves a lot of time and also helps in identifying weak passwords which can be misused within the organization and hence should be reset.

Smart automation also shines when it comes to token manipulation. Potential Security IssuesIntercepting and modifying tokens now allow testers to check for how well the application has a functioning solution against unauthorized access attempts as mentioned by (Ksiezopolski et al., 2022; Weiqi et al., 2024). You have to do a bit of detective work and see what you can get through security.

A lot of people believe that a complex password is all they need to be resistant, and you will see just how wrong this can be when using automated tools. A recent research study concluded: that approximately 80% of breaches were associated with stolen or weak passwords (Güven et al., 2021). The takeaway? Without this level of cutting-edge tech, your organization cannot take the necessary steps to secure it.

Security audits and risk assessments

Continuous monitoring of app security with tools that track suspicious access patterns and authorization changes.

Monitoring app security isn't just about ticking boxes; it's a continuous process that needs attention and the right tools. Tools that track suspicious access patterns and authorization changes can provide valuable insights into potential threats (Alshamrani et al., 2019). Regular security audits help identify vulnerabilities before they become serious problems (Dowd et al., 2006; Ying et al., 2024). For instance, a software application might have weak password policies, which could lead to unauthorized access (Harmening, 2025; Wen et al., 2023). When companies implement monitoring tools, they can quickly detect unusual activities, such as repeated failed login attempts or unexpected changes in user permissions (Rezaee et al., 2021). These tools can act like an early warning system, alerting security teams to possible breaches.

From my experience, I've seen how a company utilized automated tracking tools to catch a case where an employee's account was accessed from an unfamiliar location. This incident triggered an immediate investigation, preventing potential data loss. Moreover, risk assessments are essential for understanding the security posture of applications (Koolen et al., 2024). They help organizations prioritize resources and focus on areas that need improvement. Staying proactive in monitoring not only protects sensitive information but also fosters trust among users.

In this fast-paced digital world, where threats can emerge from unexpected places, continuous monitoring is no longer optional. It's crucial for businesses to regularly review their security measures and adapt their strategies accordingly to safeguard their data. Remember, a stitch in time saves nine! Figure 6 shows Common Authentication Failures.

Figure 6. Common Authentication Failures

6. CASE STUDIES: AUTHENTICATION AND AUTHORIZATION FAILURES IN MOBILE APPS

Case Study 1: OAuth Token Mismanagement in Third-Party Integrations

Overview of the incident involving a fitness app and vulnerabilities in token management.

This is when a party fails to obtain and store an OAuth access confidentially thus allowing third-party or eavesdroppers your credentials as shown in the mobile app fitness disaster (Andronache & Aciobănitei, 2024; Alex et al., 2022). It also discovered other key lessons, including that third-party integrations can expose potential security holes and undermine user data privacy. OAuth tokens are important for avoiding passing passwords and granting access, but when mismanaged they can become an enemy—providing attackers with valuable information. The App allowed users to connect their accounts with third-party services, but it turns out that the secure token storage was improperly implemented. By doing this, the tokens became very susceptible to being picked off by bad actors allowing them unauthorized access to sensitive data.

Now, think of a user who is diligently following his workouts and for him to come find that someone has stolen all this private data. These applications are often trusted by their users to handle health information, sometimes in a confidential manner.

Nevertheless, some applications were not properly encrypting and expiring tokens after use – which could enable attackers to compromise accounts (Vinay et al., 2024). On the other hand, a lesson to be learned from this incident is how secure coding should take precedence. For applications, we must consider token management a top priority as developers otherwise will be exposing user data.

The practical consequences of such failures extend far beyond mere technical malfunctions and dig into consumer confidence. The incident left users feeling uneasy about their data security and app engagement dived.

Case Study 2: Insecure Biometric Authentication in a Banking App

Description of the vulnerability related to unencrypted biometric data storage.

Insights into how proper assessment could have mitigated risks through encryption and layered authentication.

Mobile banking apps often promise secure access, but some vulnerabilities can compromise user data (S. Wang et al., 2024). A major concern is unencrypted biometric data storage, which can expose sensitive information like fingerprints and facial recognition data (Melzi et al., 2024). Imagine a user feeling secure while accessing their bank, only to find their biometric data hacked. The recent case of a popular banking app illustrates this risk. Hackers exploited weaknesses in the app, accessing unencrypted biometric data, and causing financial loss to users (Garrett, 2018).

Proper assessment of security measures could have mitigated these risks. For instance, implementing strong encryption for biometric data storage is essential. Encryption makes it nearly impossible for unauthorized users to access or misuse data (Gill & Devi, 2024). Layered authentication, which combines biometric verification with passwords or two-factor authentication, adds an extra level of security (Kavitha et al., 2024). Users often overlook these layers in favour of convenience, but this can lead to disaster.

The implications of insecure biometric data storage extend beyond individual users. Financial institutions risk losing customer trust and facing regulatory penalties if they fail to protect sensitive data (Kun, 2024). By emphasizing proper assessment and robust security measures, banks can foster a safer digital banking experience.

Users must advocate for their data security; demanding banks implement stronger protective measures. In a world where we often hear "your data is safe with us," it's crucial to ensure that's more than just a catchy slogan. By taking proactive steps, banks can safeguard user data while maintaining trust and integrity in the financial sector.

Case Study 3: Password Reuse Leading to Account Takeover

Examination of a social media app that suffered a breach due to users reusing passwords across platforms.

In a world where social media connects us all, it's alarming how easily it can crumble due to something as simple as password reuse (Etchells, 2024). Take, for example, a popular social media app that faced a severe breach when users repeatedly used the same passwords from other platforms. This led to unauthorized access, leaving many accounts vulnerable (Arenas et al., 2023). It's a bit like leaving your front door wide open because you trust your neighbour—only to discover they forgot to lock theirs too.

This breach exposed how many users fail to understand the risks of reusing passwords across different services. Security experts warn that this habit creates a domino effect; if one platform gets compromised, others are at risk too (Pienta et al., 2024). Imagine a hacker using stolen credentials from a data leak to waltz right into your private messages and photos. It's like a digital invasion of privacy that can lead to personal and financial harm.

The fallout wasn't just a loss of trust; it impacted users' lives. Many experienced identity theft and harassment, a reminder that our digital lives are often intertwined. Users faced the harsh reality of account takeovers, realizing that one weak password could unravel their entire online presence. This situation highlighted the urgent need for better education about password management and the use of unique passwords for each platform (Fernando, 2024).

Users must adopt stronger practices. Maybe even investing in a password manager isn't such a bad idea after all!

7. FUTURE TRENDS IN MOBILE APP AUTHENTICATION AND AUTHORIZATION

Passwordless authentication

Mobile app authentication is changing, especially with the shift towards passwordless options (Prasad, 2024). Many users find passwords annoying and hard to remember (Harisha et al., 2024). A study showed that 80% of data breaches are linked to weak passwords (Chakraborty et al., 2024). This trend has led to the rise of methods like biometrics, where users can log in using their fingerprints or faces. It's like having a key that's unique to you.

But how will we assess these new methods? Smart assessment tools are becoming more sophisticated. Instead of just checking if a password is correct, they'll analyze behaviour. For instance, they might track how quickly you type or even your location when you log in. This could help catch suspicious activity before it becomes a problem. These tools will adapt to learn from patterns, making authentication even more secure.

There's also a push for integrating artificial intelligence (AI) into these assessments. AI can predict potential threats by analyzing data in real time (Odejide & Edunjobi, 2024). As AI continues to develop, we can expect these tools to offer personalized experiences tailored to individual user habits

This new wave of authentication methods not only boosts security but also improves user experience, creating a more seamless interaction with technology. Passwordless authentication is just the beginning, and smart tools will play a crucial role in shaping a safer digital future.

Decentralized identity systems

One of the most dynamic areas of change in authentication for mobile apps is fueled by blockchain technology and decentralized identity systems. All of this makes users take control of their digital identities into their hands. During this period, there was no user with a central authority in control of credentials but a decentralized identity where individuals owned their data points (Agarkar et al., 2024).

Blockchain technology is the main lever that can contribute to improving security and privacy in mobile app authentication (Hagui, et al., 2023). This approach drastically reduces the risk of data breaches and unauthorized access by decentralizing your identity data. Blockchain smart contracts can be used to verify authorization, such that authorized users can use Gateway features (Liu et al., 2024).

This also requires sophisticated smart assessment methodologies for evolving systems designed to deal with such a decentralized manner of authentication. Assessment tools powered by blockchain allow for real-time verification of identities without relying on centralized servers. This could mean a more efficient and secure process in online learning or digital services for example.

These advances have wide-ranging implications. This in turn might well embolden users as they claw back the power of their own personal data again, waning away some of that existential dread around misuse. Users will have a better (more intuitive) experience with mobile apps (Fernando, 2024). By embracing these trends and technologies, a more secure digital experience is not only achievable but also ensures users can browse with confidence knowing their identities are secured. Figure 7 shows future trends in mobile app authentication and authorization.

Figure 7. Future Trends in Mobile App Authentication and Authorization

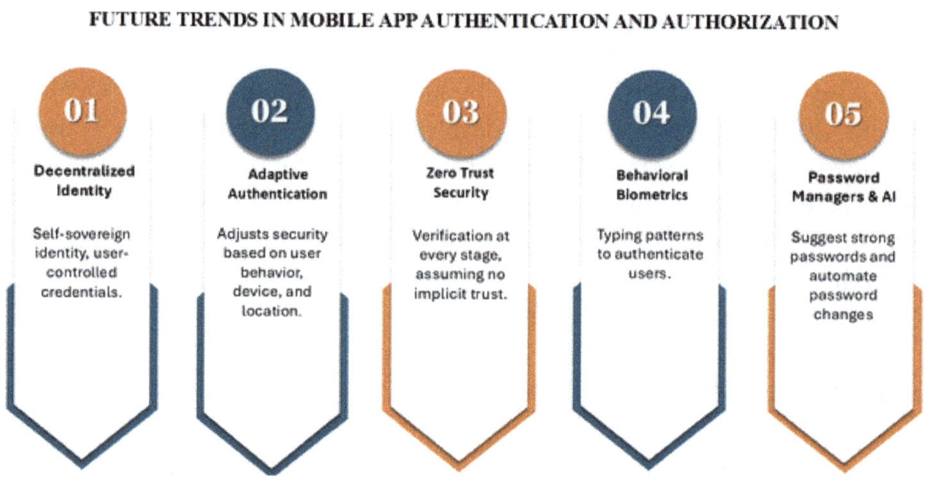

8. CONCLUSION

Recap of the critical role of robust authentication and authorization in securing mobile apps.

The importance of effective authentication and authorization in mobile apps cannot be overstated. Developers and users alike experience firsthand how gaps in these processes can leave systems vulnerable, with real-world consequences like stolen data or compromised accounts. I still recall a time when I downloaded a seemingly harmless app, only to feel uneasy about how lax its login process was—so I deleted it within hours. That moment underscored how critical it is to implement proper access controls.

We interact with apps constantly, trusting them with sensitive data like finances and health. Yet, when authentication is too simple or poorly configured, it opens the door to attacks. Similarly, inadequate authorization can expose parts of the system to users who should not have access. Effective measures, such as multi-factor authentication and role-based access control, significantly reduce risks by ensuring that only verified individuals can access appropriate sections of the app. This isn't just about technical precision; it's about creating a safer digital experience for users.

Security isn't a one-time effort, though. New methods, like biometric authentication or token-based systems, make logging in smoother but come with their challenges. As these technologies advance, tools for assessing security need to keep pace, continuously identifying weak spots before they become disasters. At the end of the day, robust access management offers peace of mind to users and ensures developers can deliver secure, reliable apps.

Importance of using smart assessment tools for continuous evaluation and strengthening of security measures.

Smart assessment tools should be continually assessing the authentication and authorization mechanisms to keep mobile apps secure. One would be forgiven for believing if the security of an app is taken care of (you have part 1 in place), then it's happy ever after. However, threats are moving — it is like dancing with developers trying to keep up all the time. This reminds me of a joke that I once heard one of my friends say: "If you close and lock the front door but forget about closing all the windows, then it doesn't matter." Enter these tools: They examine every nook and cranny; in the framework, they scan for vulnerabilities likely overlooked by developers during initial design.

The solutions do so by emulating cyber-attacks, trying out different authentication flows and live monitoring users' behaviour. They make sure that every new security release or feature addition doesn't accidentally include a whole fleet of other potential vulnerabilities. I recall a bank app that locked me out because it detected an unfamiliar login from another area. It was annoying at the time, but it illustrated how early monitoring could protect against future breaches. Although it can be a pain, this type of detection is good and helps trustworthiness by choosing security over convenience.

Continuous testing no longer serves solely to resolve issues—it drives the ability of applications to endure. Developers introduce new authentication technologies such as biometric logins or decentralized identities, and these tools evolve to measure the ease of use for users with how well they work. Ultimately, smart assessment tools aren't just technical resources; they're guardians of user trust that make sure apps stay reliable and secure far beyond launch.

Call to action for developers to integrate secure access control methods throughout the app lifecycle.

It is up to developers to build secure access control right from the first line of code and throughout an app's existence, including every post-release update. Not simply by tacking on security to an existing application, but from Day 1 the code needs to be written with security first. I have seen apps scramble to plug holes post-go-live, but often it is too little too late.

Compare that to building a house. You don't wait until the walls are up to decide where the locks go, you plan those in with security from day one. Likewise, by integrating features like MFA (Multi-factor authentication), secure token management and RBAC (Role-based access control) from the start you enable users to have a safe experience from day one. A system design that employs access control from the start or at every stage -- in testing, during deployment and afterwards is more secure than a stack added over an existing application.

This long-term focus also requires that the work does not stop after a launch. Security threats change and so should your security policies. Perform Regular Audits, Penetration Tests and User Feedback loops — to keep APPs Resilient. It could be leveraging sophisticated assessment software or establishing alerts for suspicious activity. Those developers can be the ones that are set apart — security is a journey, not just some destination. In the end, for long-term success in today's app-driven environment, someone will need to earn the trust of aimers anyway and none other than the product manager can do it.

REFERENCES

Aboukadri, S., Ouaddah, A., & Mezrioui, A. (2024). Machine learning in identity and access Management Systems: Survey and deep dive. *Computers & Security, 139*, 103729. DOI: 10.1016/j.cose.2024.103729

Agarkar, A. A., Karyakarte, M., Chavhan, G., Patil, M., Talware, R., & Kulkarni, L. (2024). Blockchain aware decentralized identity management and access control system. *Measurement. Sensors, 31*, 101032. DOI: 10.1016/j.measen.2024.101032

Aghili, S. (2024). *Leveraging blockchain technology: Governance*. Risk, Compliance, Security, and Benevolent Use Cases. DOI: 10.1201/9781003462033

Alajarmeh, N., & Ladner, R. E. (2024). Password managers use among individuals who are visually impaired: Awareness, adoption, and rejection. *International Journal of Human-Computer Interaction*, •••, 1–17. DOI: 10.1080/10447318.2024.2376356

Alassery, F., Ahmed, W., Sarraf, M., & Lawrence, V. (2015). Collision detection in wireless sensor networks through Pseudo-Coded ON-OFF pilot periods per packet: A novel Low-Complexity and Low-Power design technique. *Computer and Information Science, 8*(3). Advance online publication. DOI: 10.5539/cis.v8n3p13

Alex, S. A., Jhanjhi, N., Humayun, M., Ibrahim, A. O., & Abulfaraj, A. W. (2022). Deep LSTM Model for Diabetes Prediction with Class Balancing by SMOTE. *Electronics (Basel), 11*(17), 2737. DOI: 10.3390/electronics11172737

Alferidah, D. K., & Jhanjhi, N. (2020). Cybersecurity Impact over Bigdata and IoT Growth. *2020 International Conference on Computational Intelligence (ICCI)*. DOI: 10.1109/ICCI51257.2020.9247722

Alkinani, M. H., Almazroi, A. A., Jhanjhi, N., & Khan, N. A. (2021). 5G and IoT Based Reporting and Accident Detection (RAD) System to Deliver First Aid Box Using Unmanned Aerial Vehicle. *Sensors (Basel), 21*(20), 6905. DOI: 10.3390/s21206905 PMID: 34696118

Alothman, B., Alibrahim, O., Alenezi, N., Alhashemi, A., Alhashemi, M., Almardasi, D., Khattab, O., Joumaa, C., & Khan, M. (2023). The Development of a Secure Online System to Protect Social Networking Platforms from Security Attacks. *Applied Sciences (Basel, Switzerland), 13*(21), 11731. DOI: 10.3390/app132111731

Alrawili, R., AlQahtani, A. S., & Khan, M. K. (2024). Comprehensive survey: Biometric user authentication application, evaluation, and discussion. *Computers & Electrical Engineering, 119*, 109485. DOI: 10.1016/j.compeleceng.2024.109485

Alshamrani, A., Myneni, S., Chowdhary, A., & Huang, D. (2019). A survey on Advanced persistent threats: Techniques, solutions, challenges, and research opportunities. *IEEE Communications Surveys and Tutorials*, *21*(2), 1851–1877. DOI: 10.1109/COMST.2019.2891891

Alsmadi, I. (2023). Identity Management. In *Springer eBooks* (pp. 313–329). DOI: 10.1007/978-3-031-21651-0_12

Alzaabi, F. R., & Mehmood, A. (2024). A Review of Recent Advances, Challenges and Opportunities in Malicious Insider Threat Detection using Machine Learning Methods. *IEEE Access : Practical Innovations, Open Solutions*, *1*, 30907–30927. Advance online publication. DOI: 10.1109/ACCESS.2024.3369906

Ananna, F. F., Nowreen, R., Jahwari, S. S. R. A., Costa, E. A., Angeline, L., & Sindiramutty, S. R. (2023). Analysing Influential factors in student academic achievement: Prediction modelling and insight. *International Journal of Emerging Multidisciplinaries Computer Science & Artificial Intelligence*, *2*(1). Advance online publication. DOI: 10.54938/ijemdcsai.2023.02.1.254

Andronache, G., & Aciobănitei, L. (2024). DAnke - An Application for Promoting a Healthy Lifestyle Using Gamification and Blockchain. *2024 IEEE 18th International Symposium on Applied Computational Intelligence and Informatics (SACI)*, 000365–000370. DOI: 10.1109/SACI60582.2024.10619826

Archibong, E. E., Stephen, B. U., & Asuquo, P. (2024). Analysis of cybersecurity vulnerabilities in mobile payment applications. *Archives of Advanced Engineering Science*. DOI: 10.47852/bonviewAAES42022595

Arenas, Á., Ray, G., Hidalgo, A., & Urueña, A. (2023). How to keep your information secure? Toward a better understanding of users security behavior. *Technological Forecasting and Social Change*, *198*, 123028. DOI: 10.1016/j.techfore.2023.123028

Asatiani, A., Hakkarainen, T., Paaso, K., & Penttinen, E. (2023). Security by envelopment – a novel approach to data-security-oriented configuration of lightweight-automation systems. *European Journal of Information Systems*, •••, 1–23. DOI: 10.1080/0960085X.2023.2217362

Azam, H., Dulloo, M. I., Majeed, M. H., Wan, J. P. H., Xin, L. T., & Sindiramutty, S. R. (2023). Cybercrime Unmasked: Investigating cases and digital evidence. *International Journal of Emerging Multidisciplinaries Computer Science & Artificial Intelligence*, *2*(1). Advance online publication. DOI: 10.54938/ijemdcsai.2023.02.1.255

Azam, H., Dulloo, M. I., Majeed, M. H., Wan, J. P. H., Xin, L. T., Tajwar, M. A., & Sindiramutty, S. R. (2023). Defending the digital Frontier: IDPS and the battle against Cyber threat. *International Journal of Emerging Multidisciplinaries Computer Science & Artificial Intelligence*, *2*(1). Advance online publication. DOI: 10.54938/ijemdcsai.2023.02.1.253

Azam, H., Tajwar, M. A., Mayhialagan, S., Davis, A. J., Yik, C. J., Ali, D., & Sindiramutty, S. R. (2023). Innovations in Security: A study of cloud Computing and IoT. *International Journal of Emerging Multidisciplinaries Computer Science & Artificial Intelligence*, *2*(1). Advance online publication. DOI: 10.54938/ijemdcsai.2023.02.1.252

Azam, H., Tan, M., Pin, L. T., Syahmi, M. A., Qian, A. L. W., Jingyan, H., Uddin, M. F., & Sindiramutty, S. R. (2023). Wireless Technology Security and Privacy: A Comprehensive Study. *Preprints.org*. DOI: 10.20944/preprints202311.0664.v1

Babbar, H., Rani, S., Masud, M., Verma, S., Anand, D., & Jhanjhi, N. (2021). Load balancing algorithm for migrating switches in software-defined vehicular networks. *Computers, Materials & Continua/Computers. Computers, Materials & Continua*, *67*(1), 1301–1316. DOI: 10.32604/cmc.2021.014627

Bag, A., Patranabis, S., & Mukhopadhyay, D. (2023). CAMISE: Content Addressable Memory-Integrated Searchable Encryption. *IEEE Transactions on Circuits and Systems. I, Regular Papers*, *70*(8), 3254–3267. DOI: 10.1109/TCSI.2023.3279853

Bhatt, P., Sethi, A., Tasgaonkar, V., Shroff, J., Pendharkar, I., Desai, A., Sinha, P., Deshpande, A., Joshi, G., Rahate, A., Jain, P., Walambe, R., Kotecha, K., & Jain, N. K. (2023). Machine learning for cognitive behavioral analysis: Datasets, methods, paradigms, and research directions. *Brain Informatics*, *10*(1), 18. Advance online publication. DOI: 10.1186/s40708-023-00196-6 PMID: 37524933

Blessing, J., Hugenroth, D., Anderson, R. J., & Beresford, A. R. (2024). SOK: Web Authentication in the age of End-to-End Encryption. *arXiv (Cornell University)*. https://doi.org//arxiv.2406.18226DOI: 10.48550

Bloom, P. (2023). *The human mind: A Brief Tour of Everything We Know*. Random House.

Brohi, S. N., Jhanjhi, N., Brohi, N. N., & Brohi, M. N. (2020). Key Applications of State-of-the-Art Technologies to Mitigate and Eliminate COVID-19.pdf. *TECHRxiv*. DOI: 10.36227/techrxiv.12115596.v1

Bucko, A., Vishi, K., Krasniqi, B., & Rexha, B. (2023). Enhancing JWT authentication and authorization in web applications based on user behavior history. *Computers, 12*(4), 78. DOI: 10.3390/computers12040078

Chakraborty, S., Jackson, C., Frazier, M., & Clark, K. (2024). A Study on Password Protection and Encryption in the era of Cyber Attacks. *SoutheastCon, 2024*, 460–465. Advance online publication. DOI: 10.1109/SoutheastCon52093.2024.10500214

Chesti, I. A., Humayun, M., Sama, N. U., & Jhanjhi, N. (2020). Evolution, Mitigation, and Prevention of Ransomware. *2020 2nd International Conference on Computer and Information Sciences (ICCIS).* DOI: 10.1109/ICCIS49240.2020.9257708

Choksy, P., Chaurasia, A., Rao, U. P., & Kumar, S. (2023). Attribute based access control (ABAC) scheme with a fully flexible delegation mechanism for IoT healthcare. *Peer-to-Peer Networking and Applications, 16*(3), 1445–1467. DOI: 10.1007/s12083-023-01486-w

Devi, P. J., Dutta, M. S., Damerakonda, M., Gutti, D. N., & Domakuntla, S. K. (2023). One-Time QR-Code for Fake Product Identification. *2023 5th International Conference on Inventive Research in Computing Applications (ICIRCA)*, 583–588. DOI: 10.1109/ICIRCA57980.2023.10220784

Dogra, V., Singh, A., Verma, S., Kavita, N., Jhanjhi, N. Z., & Talib, M. N. (2021). Analyzing DistilBERT for Sentiment Classification of Banking Financial News. In *Lecture notes in networks and systems* (pp. 501–510). DOI: 10.1007/978-981-16-3153-5_53

Domoney, C. (2024). *Defending APIs: Uncover advanced defence techniques to craft secure application programming interfaces*. Packt Publishing Ltd.

Dowd, M., McDonald, J., & Schuh, J. (2006). *The art of software security assessment: Identifying and Preventing Software Vulnerabilities*. Pearson Education.

Etchells, P. (2024). *Unlocked: The Real Science of Screen Time (and how to spend it better).* Hachette UK.

Fatima-Tuz-Zahra, N., Jhanjhi, N., Brohi, S. N., Malik, N. A., & Humayun, M. (2020). Proposing a Hybrid RPL Protocol for Rank and Wormhole Attack Mitigation using Machine Learning. *2020 2nd International Conference on Computer and Information Sciences (ICCIS).* DOI: 10.1109/ICCIS49240.2020.9257607

Fernando, P. (2024). Exploring Perceptions and Habits of Sri Lankan users: A study on password management and adoption of password Managers. *International Journal of Education and Management Engineering, 14*(1), 33–40. DOI: 10.5815/ijeme.2024.01.04

Garrett, G. A. (2018). *Cybersecurity in the digital age: Tools, Techniques, & Best Practices*. Aspen Publishers.

Gill, S., & Devi, R. (2024). Enhancing Cloud Data Security using Artificial Neural Networks for Users' Account Hijacking Security Threats. *Indian Journal of Science and Technology*, *17*(34), 3538–3552. DOI: 10.17485/IJST/v17i34.2339

Gnanasekaran, S. M. (2024). *User privacy in OAUth-Based Single Sign-On systems*. DOI: 10.22215/etd/2024-15946

Gopi, R., Sathiyamoorthi, V., Selvakumar, S., Manikandan, R., Chatterjee, P., Jhanjhi, N. Z., & Luhach, A. K. (2021). Enhanced method of ANN based model for detection of DDoS attacks on multimedia internet of things. *Multimedia Tools and Applications*, *81*(19), 26739–26757. DOI: 10.1007/s11042-021-10640-6

Gouda, W., Almurafeh, M., Humayun, M., & Jhanjhi, N. Z. (2022). Detection of COVID-19 based on chest x-rays using deep learning. *Health Care*, *10*(2), 343. DOI: 10.3390/healthcare10020343 PMID: 35206957

Gupta, S., Maple, C., Crispo, B., Raja, K., Yautsiukhin, A., & Martinelli, F. (2023a). A survey of human-computer interaction (HCI) & natural habits-based behavioural biometric modalities for user recognition schemes. *Pattern Recognition*, *139*, 109453. DOI: 10.1016/j.patcog.2023.109453

Güven, E. Y., Boyaci, A., & Aydin, M. A. (2021). A novel password policy focusing on altering user password selection habits: A statistical analysis on breached data. *Computers & Security*, *113*, 102560. DOI: 10.1016/j.cose.2021.102560

Hagui, I., Msolli, A., Henda, N. B., Helali, A., Gassoumi, A., Nguyen, T. P., & Hassen, F. (2023). A blockchain-based security system with light cryptography for user authentication security. *Multimedia Tools and Applications*, *83*(17), 52451–52480. DOI: 10.1007/s11042-023-17643-5

Harisha, N., Naik, S. R., Vasudeva, S. S., Shrilakshmi, K., & Kothwal, V. (2024). Advancements in User Security: Enhancing Usability with Graphical Password Authentication. *2024 2nd International Conference on Intelligent Data Communication Technologies and Internet of Things (IDCIoT)*. DOI: 10.1109/IDCIoT59759.2024.10467993

Harmening, J. (2025). Security Management Systems. In *Elsevier eBooks* (pp. 433–439). DOI: 10.1016/B978-0-443-13223-0.00025-4

He, D., Liu, Z., Zhu, S., Chan, S., & Guizani, M. (2024). Special characters usage and its effect on password security. *IEEE Internet of Things Journal*, *11*(11), 19440–19453. DOI: 10.1109/JIOT.2024.3367323

Humayun, M., Sujatha, R., Almuayqil, S. N., & Jhanjhi, N. Z. (2022). A Transfer Learning Approach with a Convolutional Neural Network for the Classification of Lung Carcinoma. *Health Care, 10*(6), 1058. DOI: 10.3390/healthcare10061058 PMID: 35742109

Hussain, K., Rahmatyar, A. R., Riskhan, B., Sheikh, M. U., & Sindiramutty, S. R. (2024). Threats and Vulnerabilities of Wireless Networks in the Internet of Things (IoT). *2024 IEEE 1st Karachi Section Humanitarian Technology Conference (KHI-HTC), 2*, 1–8. DOI: 10.1109/KHI-HTC60760.2024.10482197

Hwaitat, A. K. A., Fakhouri, H. N., Alawida, M., Atoum, M. S., Abu-Salih, B., Salah, I. K. M., Al-Sharaeh, S., & Alassaf, N. (2024). Overview of mobile attack detection and prevention techniques using Machine Learning. [iJIM]. *International Journal of Interactive Mobile Technologies, 18*(10), 125–157. DOI: 10.3991/ijim.v18i10.46485

Igwenagu, U. T. I., Salami, A. A., Arigbabu, A. S., Mesode, C. E., Oladoyinbo, T. O., & Olaniyi, O. O. (2024). Securing the digital frontier: Strategies for cloud computing security, database protection, and comprehensive penetration testing. *Journal of Engineering Research and Reports, 26*(6), 60–75. DOI: 10.9734/jerr/2024/v26i61162

Jamwal, S., Cano, J., Lee, G. M., Tran, N. H., & Truong, N. (2024). A survey on ethereum pseudonymity: Techniques, challenges, and future directions. *Journal of Network and Computer Applications, 232*, 104019. DOI: 10.1016/j.jnca.2024.104019

Jhanjhi, N., Humayun, M., & Almuayqil, S. N. (2021). Cyber security and privacy issues in industrial internet of things. *Computer Systems Science and Engineering, 37*(3), 361–380. DOI: 10.32604/csse.2021.015206

Jiang, X., Dou, R., He, Q., Zhang, X., & Dou, W. (2023). EdgeAuth: An intelligent token-based collaborative authentication scheme. *Software, Practice & Experience, 54*(10), 2007–2024. DOI: 10.1002/spe.3206

Josyula, H. P. (2023). Fraud detection in fintech leveraging machine learning and behavioral analytics. *Research Square (Research Square)*. DOI: 10.21203/rs.3.rs-3548343/v1

Jun, A. Y. M., Jinu, B. A., Seng, L. K., Maharaiq, M. H. F. B. Z., Khongsuwan, W., Junn, B. T. K., Hao, A. W., & Sindiramutty, S. R. (2024). Exploring the Impact of Crypto-Ransomware on Critical Industries: Case Studies and Solutions. *Preprints.org*. DOI: 10.20944/preprints202409.1325.v1

Kammoun, A., Slama, R., Tabia, H., Ouni, T., & Abid, M. (2022a). Generative Adversarial Networks for face generation: A survey. *ACM Computing Surveys*, 1122445.1122456. Advance online publication. DOI: 10.1145/1122445.1122456

Kanungo, K., Khatoliya, R., Arora, V., Bari, A., Bhattacharya, A., Maity, M., & Chakravarty, S. (2024). How Many Hands in the Cookie Jar? Examining Privacy Implications of Popular Apps in India. *2024 IEEE 9th European Symposium on Security and Privacy (EuroS&P)*, 741–757. DOI: 10.1109/EuroSP60621.2024.00046

Karim, N. A., Kanaker, H., Abdulraheem, W. K., Ghaith, M. A., Alhroob, E., & Alali, A. M. F. (2024). Choosing the right MFA method for online systems: A comparative analysis. *International Journal of Data and Network Science*, *8*(1), 201–212. DOI: 10.5267/j.ijdns.2023.10.003

Kaur, A. A., & Mustafa, K. K. (2019). A Critical appraisal on Password based Authentication. *International Journal of Computer Network and Information Security*, *11*(1), 47–61. DOI: 10.5815/ijcnis.2019.01.05

Kaushik, K., Ouaissa, M., & Chaudhary, A. (2024). *Advanced techniques and applications of cybersecurity and forensics*. CRC Press. DOI: 10.1201/9781003386926

Kavitha, G., Prasannakumar, V., & Pranav, R. (2024). Enhancing Digital Security: A Comprehensive Multi-Model Authentication Framework Leveraging Cryptography and Biometrics. *2024 8th International Conference on Inventive Systems and Control (ICISC)*, 476–486. DOI: 10.1109/ICISC62624.2024.00086

Kävrestad, J., Birath, M., & Clarke, N. (2024). Decryption and password cracking. In *Texts in computer science* (pp. 43–53). DOI: 10.1007/978-3-031-53649-6_5

Khamisa, A. (2021). Token economies. In *Emerald Publishing Limited eBooks* (pp. 121–140). DOI: 10.1108/978-1-83982-198-120211011

Khan, J. A. (2024). Role-Based access Control (RBAC) and Attribute-Based Access Control (ABAC). In *Advances in information security, privacy, and ethics book series* (pp. 113–126). DOI: 10.4018/979-8-3693-1431-9.ch005

Khan, S., Devlen, C., Manno, M., & Hou, D. (2024). Mouse Dynamics Behavioral Biometrics: A survey. *ACM Computing Surveys*, *56*(6), 1–33. DOI: 10.1145/3640311

Kim, H., Kim, Y., & Kim, S. (2024). A Study on the Security Requirements Analysis to build a Zero Trust-based Remote Work Environment. *arXiv (Cornell University)*. https://doi.org//arxiv.2401.03675DOI: 10.48550

Kissell, J. (2024). Take Control of Your Passwords, *4th Edition*. alt concepts.

Kiyani, F. F., Hamid, B., Humayun, M., Sindiramutty, S. R. L., & Chowdhury, S. (2024). Discovery of Influential Publications Using Research Article's Usage Context. *2024 International Conference on Emerging Trends in Networks and Computer Communications (ETNCC)*, 1–7. DOI: 10.1109/ETNCC63262.2024.10767576

Kizza, J. M. (2024a). Access control and authorization. In *Texts in computer science* (pp. 195–214). DOI: 10.1007/978-3-031-47549-8_9

Kizza, J. M. (2024b). Authentication. In *Texts in computer science* (pp. 215–238). DOI: 10.1007/978-3-031-47549-8_10

Koolen, C., Wuyts, K., Joosen, W., & Valcke, P. (2024). From insight to compliance: Appropriate technical and organisational security measures through the lens of cybersecurity maturity models. *Computer Law & Security Review*, *52*, 105914. DOI: 10.1016/j.clsr.2023.105914

Ksiezopolski, B., Mazur, K., Miskiewicz, M., & Rusinek, D. (2022). Teaching a Hands-On CTF-Based Web Application Security course. *Electronics (Basel)*, *11*(21), 3517. DOI: 10.3390/electronics11213517

Kumar, M. S., Vimal, S., Jhanjhi, N., Dhanabalan, S. S., & Alhumyani, H. A. (2021). Blockchain based peer to peer communication in autonomous drone operation. *Energy Reports*, *7*, 7925–7939. DOI: 10.1016/j.egyr.2021.08.073

Kun, E. (2024). Challenges in regulating cloud service providers in EU financial regulation: From operational to systemic risks, and examining challenges of the new oversight regime for critical cloud service providers under the Digital Operational Resilience Act. *Computer Law & Security Review*, *52*, 105931. DOI: 10.1016/j.clsr.2023.105931

Lee, M. (2024). *Hacks, leaks, and revelations: The Art of Analyzing Hacked and Leaked Data*. No Starch Press.

Lim, M., Abdullah, A., Jhanjhi, N., Khan, M. K., & Supramaniam, M. (2019). Link Prediction in Time-Evolving Criminal Network with deep Reinforcement learning technique. *IEEE Access : Practical Innovations, Open Solutions*, *7*, 184797–184807. DOI: 10.1109/ACCESS.2019.2958873

Linqiang, Y., Sindiramutty, S. R. L., Ashraf, H., Muzammal, S. M., Balakrishnan, S. P., Gupta, S., & Kavita, N. (2024). Intelligent Household Waste Classification System Based on Machine Learning. *2024 International Conference on Emerging Trends in Networks and Computer Communications (ETNCC)*, 760–768. DOI: 10.1109/ETNCC63262.2024.10767563

Liu, Y., He, J., Li, X., Chen, J., Liu, X., Peng, S., Cao, H., & Wang, Y. (2024). An overview of blockchain smart contract execution mechanism. *Journal of Industrial Information Integration*, *41*, 100674. DOI: 10.1016/j.jii.2024.100674

Lykousas, N., & Patsakis, C. (2024). Decoding developer password patterns: A comparative analysis of password extraction and selection practices. *Computers & Security*, *145*, 103974. DOI: 10.1016/j.cose.2024.103974

Mali, S. (2024). Assessing the effectiveness of Multi-Factor authentication in Cloud-Based big data environments. *Internet of Things and Cloud Computing*, *12*(2), 17–27. DOI: 10.11648/j.iotcc.20241202.11

Manchuri, A., Kakera, A., Saleh, A., & Raja, S. (2024). pplication of Supervised Machine Learning Models in Biodiesel Production Research - A Short Review. *Borneo Journal of Sciences and Technology*. DOI: 10.35370/bjost.2024.6.1-10

Melzi, P., Rathgeb, C., Tolosana, R., Vera, R., & Busch, C. (2024). An Overview of Privacy-Enhancing Technologies in biometric Recognition. *ACM Computing Surveys*, *56*(12), 1–28. Advance online publication. DOI: 10.1145/3664596

Najat, R., Haitam, E., & Jaafar, A. (2023). Comparative study of the Security Analysis of IoT systems using attack trees algorithm. *E3S Web of Conferences, 412*, 01087. DOI: 10.1051/e3sconf/202341201087

Nanda, A., Jeong, J. J., Shah, S. W. A., Nosouhi, M., & Doss, R. (2023). Examining usable security features and user perceptions of physical authentication devices. *Computers & Security*, *139*, 103664. DOI: 10.1016/j.cose.2023.103664

Nayyar, A., Gadhavi, L., & Zaman, N. (2021). Machine learning in healthcare: review, opportunities and challenges. In *Elsevier eBooks* (pp. 23–45). DOI: 10.1016/B978-0-12-821229-5.00011-2

Noble, B. D., & Corner, M. D. (2002). *The case for transient authentication*. ACM., DOI: 10.1145/1133373.1133377

Odejide, N. O. A., & Edunjobi, N. T. E. (2024). AI IN PROJECT MANAGEMENT: EXPLORING THEORETICAL MODELS FOR DECISION-MAKING AND RISK MANAGEMENT. *Engineering Science & Technology Journal*, *5*(3), 1072–1085. DOI: 10.51594/estj.v5i3.959

Olabanji, S. O., Olaniyi, O. O., Adigwe, C. S., Okunleye, O. J., & Oladoyinbo, T. O. (2024). AI for Identity and Access Management (IAM) in the Cloud: Exploring the Potential of Artificial Intelligence to Improve User Authentication, Authorization, and Access Control within Cloud-Based Systems. SSRN *Electronic Journal*. DOI: 10.2139/ssrn.4706726

Oliveira, J., Santin, A., Viegas, E., & Horchulhack, P. (2024). A non-interactive One-Time Password-Based method to enhance the vault security. In *Lecture notes on data engineering and communications technologies* (pp. 201–213). DOI: 10.1007/978-3-031-57916-5_18

Paudel, R., & Al-Ameen, M. N. (2024). Priming through Persuasion: Towards Secure Password Behavior. *Proceedings of the ACM on Human-Computer Interaction, 8*(CSCW1), 1–27. DOI: 10.1145/3637387

Pham, H., Nhuan, N. D. O., & Huynh-Tuong, N. (2023). A Fine-grained Access Control Model with Enhanced Flexibility and On-chain Policy Execution for IoT Systems. *International Journal of Advanced Computer Science and Applications, 14*(6). Advance online publication. DOI: 10.14569/IJACSA.2023.0140609

Pienta, D., Thatcher, J. B., Wright, R. T., & Roth, P. L. (2024). An empirical investigation of the unintended consequences of vulnerability assessments leading to betrayal. *Journal of the Association for Information Systems, 25*(4), 1079–1116. DOI: 10.17705/1jais.00875

Prasad, A. (2024). A Comparative Study of Passwordless Authentication. *TechRxiv*. DOI: 10.36227/techrxiv.171560547.71979752/v1

Prinakaa, S., Bavanika, V., Sanjana, S., Srinivasan, S., & Sarasvathi, V. (2024, April). A Real-Time Approach to Detecting API Abuses Based on Behavioral Patterns. In 2024 8th International Conference on Cryptography, Security and Privacy (CSP) (pp. 24-28). IEEE.

Prümmer, J. (2024). The Role of Cognition in Developing Successful Cybersecurity Training Programs – Passive vs. Active Engagement. In *Lecture notes in computer science* (pp. 185–199). DOI: 10.1007/978-3-031-61572-6_13

Rabiah, A. B., Alsoliman, A., Shashwat, Y., Richelson, S., & Abu-Ghazaleh, N. (2024). Token-based Vehicular Security System (TVSS): scalable, secure, low-latency public key infrastructure for connected vehicles. *arXiv (Cornell University)*. https://doi.org//arxiv.2402.18365DOI: 10.48550

Raphael, M. M., Kanta, A., Seebonn, R., Dürmuth, M., & Cobb, C. (2024). *Batman Hacked my Password: A {Subtitle-Based} analysis of password depiction in movies*. https://www.usenix.org/conference/soups2024/presentation/raphael

Ravichandran, N., Tewaraja, T., Rajasegaran, V., Kumar, S. S., Gunasekar, S. K. L., & Sindiramutty, S. R. (2024). Comprehensive Review Analysis and Countermeasures for Cybersecurity Threats: DDoS, Ransomware, and Trojan Horse Attacks. *Preprints.org*. DOI: 10.20944/preprints202409.1369.v1

Rezaee, K., Rezakhani, S. M., Khosravi, M. R., & Moghimi, M. K. (2021). A survey on deep learning-based real-time crowd anomaly detection for secure distributed video surveillance. *Personal and Ubiquitous Computing*. Advance online publication. DOI: 10.1007/s00779-021-01586-5

Rooney, M. J., Levy, Y., Li, W., & Kumar, A. (2024). Comparing experts' and users' perspectives on the use of password workarounds and the risk of data breaches. *Information and Computer Security*. Advance online publication. DOI: 10.1108/ICS-05-2024-0116

Saadi, Z. M., Sadiq, A. T., Akif, O. Z., & Farhan, A. K. (2024a). A Survey: Security Vulnerabilities and Protective Strategies for graphical Passwords. *Electronics (Basel), 13*(15), 3042. DOI: 10.3390/electronics13153042

Saadi, Z. M., Sadiq, A. T., Akif, O. Z., & Farhan, A. K. (2024b). A Survey: Security Vulnerabilities and Protective Strategies for graphical Passwords. *Electronics (Basel), 13*(15), 3042. DOI: 10.3390/electronics13153042

Salem, A., Sharieh, A., & Jabri, R. (2022). Online user authentication System using keystroke dynamics. *Journal of Computer Security, 31*(3), 185–215. DOI: 10.3233/JCS-210081

Saqib, M., & Moon, A. H. (2022). A systematic security assessment and review of internet of things in the context of authentication. *Computers & Security, 125*, 103053. DOI: 10.1016/j.cose.2022.103053

Sarfaraz, A., Chakrabortty, R. K., & Essam, D. L. (2023). AccessChain: An access control framework to protect data access in blockchain enabled supply chain. *Future Generation Computer Systems, 148*, 380–394. DOI: 10.1016/j.future.2023.06.009

Saxena, U. R., & Alam, T. (2023). Role-based access using partial homomorphic encryption for securing cloud data. *International Journal of System Assurance Engineering and Management, 14*(3), 950–966. DOI: 10.1007/s13198-023-01896-2

Seng, Y. J., Cen, T. Y., Raslan, M. H. B. M., Subramaniam, M. R., Xin, L. Y., Kin, S. J., Long, M. S., & Sindiramutty, S. R. (2024). In-Depth Analysis and Countermeasures for Ransomware Attacks: Case Studies and Recommendations. *Preprints.org*. DOI: 10.20944/preprints202408.2261.v1

Shah, I. A., Jhanjhi, N. Z., & Laraib, A. (2022). Cybersecurity and blockchain usage in contemporary business. In *Advances in information security, privacy, and ethics book series* (pp. 49–64). DOI: 10.4018/978-1-6684-5284-4.ch003

Sharma, R., Singh, A., Kavita, N., Jhanjhi, N. Z., Masud, M., Jaha, E. S., & Verma, S. (2021). Plant disease diagnosis and image classification using deep learning. *Computers, Materials & Continua/Computers. Materials & Continua (Print)*, *71*(2), 2125–2140. DOI: 10.32604/cmc.2022.020017

Sheng, L. T., Azman, A. B., Khan, N. A., Sindiramutty, S. R., Singh, S., & Verma, S. (2024). Reinforcement learning-based traffic signal control. *2024 International Conference on Emerging Trends in Networks and Computer Communications (ETNCC)*, 792–798. DOI: 10.1109/ETNCC63262.2024.10767501

Sindiramutty, S. R. (2024). Autonomous Threat Hunting: a future paradigm for AI-Driven Threat intelligence. *arXiv (Cornell University)*. https://doi.org//arxiv.2401.00286DOI: 10.48550

Sindiramutty, S. R., Prabagaran, K. R. V., Akbar, R., Hussain, M., & Malik, N. A. (2024). Overview of Generative AI techniques for cybersecurity. In *Advances in information security, privacy, and ethics book series* (pp. 1–52). DOI: 10.4018/979-8-3693-5415-5.ch001

Sindiramutty, S. R., Prabagaran, K. R. V., Jhanjhi, N. Z., Ghazanfar, M. A., Malik, N. A., & Soomro, T. R. (2024). Security Considerations in Generative AI for web Applications. In *Advances in information security, privacy, and ethics book series* (pp. 281–332). DOI: 10.4018/979-8-3693-5415-5.ch009

Sindiramutty, S. R., Prabagaran, K. R. V., Jhanjhi, N. Z., Murugesan, R. K., Brohi, S. N., & Masud, M. (2024). Generative AI in network security and intrusion detection. In *Advances in information security, privacy, and ethics book series* (pp. 77–124). DOI: 10.4018/979-8-3693-5415-5.ch003

Sindiramutty, S. R., Prabagaran, K. R. V., Jhanjhi, N. Z., Murugesan, R. K., Brohi, S. N., & Wei, G. W. (2024). Generative AI for threat intelligence and information sharing. In *Advances in digital crime, forensics, and cyber terrorism book series* (pp. 191–234). DOI: 10.4018/979-8-3693-8944-7.ch006

Sindiramutty, S. R., Prabagaran, K. R. V., Jhanjhi, N. Z., Murugesan, R. K., Malik, N. A., & Hussain, M. (2024). Ethics and transparency in secure web model generation. In *Advances in information security, privacy, and ethics book series* (pp. 411–464). DOI: 10.4018/979-8-3693-5415-5.ch012

Singhal, V., Jain, S. S., Anand, D., Singh, A., Verma, S., Kavita, N., Rodrigues, J. J. P. C., Jhanjhi, N. Z., Ghosh, U., Jo, O., & Iwendi, C. (2020). Artificial Intelligence Enabled Road Vehicle-Train Collision Risk Assessment Framework for Unmanned railway level crossings. *IEEE Access : Practical Innovations, Open Solutions*, *8*, 113790–113806. DOI: 10.1109/ACCESS.2020.3002416

Trampert, L., Stock, B., & Roth, S. (2023). *Honey, I Cached our Security Tokens Re-usage of Security Tokens in the Wild.* ACM., DOI: 10.1145/3607199.3607223

Upadrista, V., Nazir, S., & Tianfield, H. (2023). Secure data sharing with blockchain for remote health monitoring applications: A review. *Journal of Reliable Intelligent Environments*, *9*(3), 349–368. DOI: 10.1007/s40860-023-00204-w PMID: 37359293

Ur, B., Bees, J., Segreti, S. M., Bauer, L., Christin, N., & Cranor, L. F. (2016). *Do Users' Perceptions of Password Security Match Reality?* ACM., DOI: 10.1145/2858036.2858546

Vinay, M. D., Kumar, M. H., Hemanth, B., & Tomar, D. S. (2024). The Demonstration of Android Vulnerabilities. *2024 IEEE International Students' Conference on Electrical, Electronics and Computer Science (SCEECS), 11*, 1–10. DOI: 10.1109/SCEECS61402.2024.10482193

Waheed, A., Seegolam, B., Jowaheer, M. F., Sze, C. L. X., Hua, E. T. F., & Sindiramutty, S. R. (2024). Zero-Day Exploits in Cybersecurity: Case Studies and Countermeasure. *preprints.org*. DOI: 10.20944/preprints202407.2338.v1

Wang, L., Liu, X., Lei, T., Song, W., Guo, S., & Ren, P. (2024). Security research for Android remote assistance apps. In *Lecture notes in computer science* (pp. 256–276). DOI: 10.1007/978-981-97-5101-3_14

Wang, S., Asif, M., Shahzad, M. F., & Ashfaq, M. (2024). Data privacy and cybersecurity challenges in the digital transformation of the banking sector. *Computers & Security*, *147*, 104051. DOI: 10.1016/j.cose.2024.104051

Wear, S. (2018). *Burp Suite Cookbook: Practical recipes to help you master web penetration testing with Burp Suite*. Packt Publishing Ltd.

Wee, A. K., Chekole, E. G., & Zhou, J. (2024). Excavating Vulnerabilities lurking in Multi-Factor Authentication Protocols: A Systematic Security analysis. *arXiv (Cornell University)*. https://doi.org//arxiv.2407.20459DOI: 10.48550

Weiqi, X., Hooi, S. T. C., Sindiramutty, S. R. L., Asirvatham, D. L., Kumar, D., & Verma, S. (2024). Surface Anomaly Detection Using Machine Learning Technique. *2024 International Conference on Emerging Trends in Networks and Computer Communications (ETNCC)*, 1–7. DOI: 10.1109/ETNCC63262.2024.10767562

Wen, B. O. T., Syahriza, N., Xian, N. C. W., Wei, N. G., Shen, T. Z., Hin, Y. Z., Sindiramutty, S. R., & Nicole, T. Y. F. (2023). Detecting cyber threats with a Graph-Based NIDPS. In *Advances in logistics, operations, and management science book series* (pp. 36–74). DOI: 10.4018/978-1-6684-7625-3.ch002

Yang, Y., Wang, J., Liu, P., Fu, A., & Zhang, Y. (2024). Uncovering access token Security flaws in Multi-User scenario of Smart Home Platforms. *IEEE Internet of Things Journal, 1*(22), 36841–36857. Advance online publication. DOI: 10.1109/JIOT.2024.3429417

Yang, Y., Zhao, J., Li, Y., Li, X., Liu, P., & Zhang, Y. (2024). Sharing can be Threatening: Uncovering Security Flaws of RBAC Model on Smart Home Platforms. *IEEE Transactions on Dependable and Secure Computing*, •••, 1–18. DOI: 10.1109/TDSC.2024.3369634

Ying, X., Murugesan, R. K., Sindiramutty, S. R., Wei, G. W., Balakrishnan, S., Kumar, D., & Verma, S. (2024). Scene Text Recognition using Deep Learning Techniques. *2024 International Conference on Emerging Trends in Networks and Computer Communications (ETNCC)*, 1–9. DOI: 10.1109/ETNCC63262.2024.10767484

Zhang, Z., Chen, H., Huang, R., Zhu, L., Ma, S., Leifer, L., & Liu, W. (2024). Automated Classification of user needs for beginner user experience designers: A Kano model and text analysis approach using deep learning. *AI, 5*(1), 364–382. DOI: 10.3390/ai5010018

Zhou, W., Jia, Y., Peng, A., Zhang, Y., & Liu, P. (2018). The effect of IoT new features on security and privacy: New threats, existing solutions, and challenges yet to be solved. *IEEE Internet of Things Journal, 6*(2), 1606–1616. DOI: 10.1109/JIOT.2018.2847733

Zirillo, M., Aviza, S., Cruz, M., Lee, H., LaMalva, G., Galvez, A., Dragos, D., & Schmeelk, S. (2024). Examining the privacy and security of mobile refugee applications. In *Lecture notes in networks and systems* (pp. 410–427). DOI: 10.1007/978-3-031-62273-1_26

Chapter 4
Revolutionizing Healthcare With Cloud Computing: Opportunities, Challenges, and Innovations for Secure Patient-Centered Systems

Ravikumar Ch
https://orcid.org/0000-0003-0809-5545
Sreenidhi University, Hyderabad, India

Burri Naresh
CVR College of Engineering, India

Isha Batra
Lovely Professional University, India

Arun Malik
Lovely Professional University, India

Padmaja Grandhe
Sreenidhi University, India

ABSTRACT

Cloud computing is revolutionizing healthcare by improving service quality, efficiency, and affordability, while also addressing the evolving challenges in patient care. With the rapid digital transformation in healthcare, cloud-based systems offer new opportunities for data storage, patient monitoring, and healthcare management.

DOI: 10.4018/979-8-3693-6925-8.ch004

However, as more sensitive medical data moves to the cloud, security concerns regarding data privacy and third-party remote storage become critical issues. This research examines the potential of healthcare cloud computing, highlighting its benefits and challenges, including security and legal considerations. The paper introduces TDBNN, a deep learning-based system designed to enhance the security of cloud-stored healthcare data, ensuring better protection of patient privacy. This work provides insights into the future of cloud-based healthcare systems and their implications for patient-centered care.

1. INTRODUCTION

Healthcare is a constantly looked upon and ever more broadly defined foundation of social comfort, whose main concerns revolved around the promotion, observation and preservation of the health of people. The incorporation of the cloud computing networks in the healthcare systems has in the past years led to many innovations to enhance the quality and accessibility of the healthcare services. The world healthcare market is changing fast with an emergence in demands of cost-effective, efficient, and high quality healthcare services (Boudlal, Serrhini, & Tahiri, 2022). One of such technologies that have taken important space in this sector is cloud computing which has helped to improve healthcare delivery to be smarter and efficient. These new developments can be aimed as bettering the quality of healthcare services as well as minimizing costs and patient misery, providing an easier and patient-centered healthcare society (Sobhy, El-Sonbaty, & Abou Elnasr, 2012).

Although the benefits of the cloud computing in the healthcare are considerable, its implementation comes with a challenge, especially in the safety and lawfulness of data. One of the most serious issues is safety of sensitive information about the patients kept remotely by third-party vendors. The risk of data leaks of malicious use of such data is quite serious, which is why the security question lies at the heart of mass adoption of cloud technologies in the medical field (Agapito & Cannataro, 2023). In this way, the proposed study is supposed to describe the opportunities and difficulties connected with cloud computing in healthcare with the references to the security aspect and the ways of dealing with the challenges. This paper can help the existing debate on the role of cloud computing in changing the healthcare system by analyzing the existing case studies and offering practical insights.

The healthcare industry has experienced accelerated growth of technologies, which are based on clouds and which have increased efficiency, flexibility, and scale of applications. Nevertheless, with the transfer of medical data to the cloud, keeping private data of sensitive patients a priority is underway. The proposed paper will discuss the implementation of cloud computing in the health sector, and the new model has been proposed to enable cloud healthcare strengthen cybersecurity, the Termite Deep Belief Neural Network (TDBNN), is meant to improve cybersecurity of cloud-based health organizations, especially as part of smart cities (Balasubramaniam et al., 2023). The study will also point at how fog computing is responding to such important concerns like latency and access to resources in health applications, enhancing the general management process of healthcare operations and lowering expenses to run operations (Mutlag et al., 2021).

Figure 1. Cloud Computing Integration in Healthcare Systems

The figure-1 illustration shows how cloud computing is converting the future of healthcare delivery through its three main advantages access, cost reduction and data security. It also states the major pitfalls of implementation such as privacy issues, reliability and latency of the system. The visualization offers a succinct yet concise summary off the opportunities and risks that are related to the adoption of cloud technologies in healthcare.

2. LITERATURE REVIEW

Healthcare has completely changed due to the introduction of cloud computing because it streamlines the storage, management, and dissemination of healthcare information. In their article, Boudlal, Serrhini, and Tahiri (2022) explain that the use of cloud computing in the healthcare sector has allowed them to ensure the proper storage as well as real time accessibility to the information about patients, and as such, they were able to enhance healthcare service delivery. Such healthcare systems based in the cloud have been implemented to most healthcare areas, blood pressure, heartbeat and ECG control and have improved the access and quality of the health information made available to the health practitioners and the patients themselves (Sobhy, El-Sonbaty, & Abou Elnasr, 2012). Although the benefits of using cloud computing in healthcare are wide-spreading, there have emerged new challenges mostly covering data security and privacy, which are vital areas in healthcare (Agapito & Cannataro, 2023).

With the development of cloud computing there have developed newer technologies integrating with some limitations of the cloud services like fog computing. A reduction in latency and an improved real-time feature of healthcare systems may be achieved with the help of fog computing, which processes information near the source. Mutlag et al. (2021) have proposed the Critical Healthcare Task Management (CHTM) model that has utilized multi-agent systems in the fog-cloud computing platform. The model eliminates resource waste, lessens network delays and enhances energy efficiency, showing how fog computing can be used to determine real-time healthcare requirements. CHTM framework is able to cut down network occupancy, response time, latency, energy consumption, and costs to as much as 90 percent lower than the traditional cloud-based systems (Mutlag et al., 2021).

It is also because of cloud computing that Internet of Things (IoT) technologies are integrated into the healthcare system. Nasser et al. (2021) provided a smart healthcare framework with the combination of cloud computing and IoT to track and observe COVID-19, which demonstrates a promising future of cloud technologies in patient remote monitoring. Cloud computing has become indispensable in patient healthcare remote management through the ability to track the vital signs of patients, including oxygen saturation and temperature, in real-time in the context of the COVID-19 epidemic when it was impossible to be in contact with patients physically. Similarly, Hennebelle, Materwala, and Ismail (2023) investigated how machine learning-based cloud systems could forecast the development of type 2 diabetes and presented the value of cloud computing and AI in preventing and intervening at an early stage.

Security and privacy are also major issues in the future of cloud computing in healthcare, and they pose a tremendous impediment to widespread implementation. The solution of cloud-based administration of Electronic Health Records (EHR) triggers some fears regarding the risk of unauthorized accesses and data losses. To solve such issues, a solution has been suggested in the form of blockchain technology so that patient information can be secure and authentic. Amanat et al. (2022) have written about the issue of the blockchain-enhanced security of EHRs, which offers a decentralized, transparent, and tamper-proof structure regarding how the data is stored and shared. The blockchain technology can provide a potentially solution in ensuring the safety of the transfer of the patient data in reducing the risk of cloud computing involved in healthcare (Rani et al., 2022).

In addition to its security purposes, cloud computing in the sphere of healthcare could be referred to the optimization of healthcare processes. Onyebuchi et al. (2022) focused on the ways cloud-based data warehouse can facilitate new e-healthcare systems (on the example of automation and optimization of patient treatment schedules, enhancing the efficiency of the overall healthcare system). Nonetheless, the issue of secure and effective management of data is present. According to Zala et al. (2022), security of privacy in cloud systems is imperative because third-party services should have powerful security solutions to block the exposure of sensitive patient information to various actors.

The other opportunity, or rather problem of cloud computing in the medical sector is the accuracy of the diagnostic models. Hennebelle et al. (2023) made it clear that machine learning models have significantly increased the level of diagnostic accuracy, but human experience is still pivotal in healthcare sector especially during diagnosis and treatment. The excessive use of AI and machine learning without taking into account the specifics of a patient may lead to mistaken disease diagnosis drawing attention to the necessity of the human control in medicine decision-making process (Somani & Trivedi, 2019). Also, the misuse of healthcare data may occur during its transmission between various systems, which implies that the violation of privacy may be most relevant in cloud-based healthcare systems (Griebel et al., 2015).

The sharing and storage of healthcare information can be resolved using Blockchain-based strategies as indicated by the studies conducted by Amanat et al. (2022) and Rani et al. (2022). Such systems apply cryptographic methods to guarantee data integrity and confidentiality of information about patients, which is secure and one that cannot be tampered with. Integrating blockchain with cloud computing enables the secure storage and sharing of patient data by healthcare organizations, avoiding all risks associated with unauthorized access and a breach of information.

Moreover, Elhadad, Alanazi, Taloba, and Abozeid (2022) studied how to incorporate the fog computing in healthcare monitoring systems to improve the real-time data processing of healthcare apps. With the help of fog computing, together with cloud technologies, the effectiveness of the healthcare system can be enhanced due to the lower latency and better resource distribution, which is particularly relevant in time-demanding processes such as patient monitoring and emergency management.

To sum up, the literature proves the disruptive power of cloud computing in healthcare, especially in better data management, accessibility, and security. Nonetheless, issues of privacy, data safety, and interoperability of systems should be overcome to reach the full potential of cloud technologies. Correlating blockchain, IoT, and fog computing with cloud systems contain potential solutions to these problems and will improve the overall healthcare provision.

3. METHODOLOGY

The study seeks to investigate issues, safety measures and the dynamic nature of cloud computing in the healthcare sector and, in particular, the effective prediction and allocation of resources in a fog computing environment. The aim is to find out the existing barriers to integration, the available opportunities, and how best to apply cloud technologies to enhance delivery and management of healthcare. This was initiated by a thorough literature review in order to identify extant trends, challenges and opportunities in the merging of cloud computing and healthcare systems.

In order to direct the research a number of key research questions were developed to direct the research as follows:

a) What can be done to efficiently use cloud computing in enhancing healthcare systems?
b) What are the main issues and risks of adopting cloud computing to healthcare settings?
c) What are the healthcare applications that would gain most with the implementation of cloud computing?
d) What does it take in terms of security to implement and maintain the cloud-based healthcare systems?

The initial procedure of the methodology was the search of the academic databases of appropriated literature. PubMed is the major database that was used to find articles related to the two concepts of cloud computing and healthcare. The queries were written so that they may solve the research questions and hence the retrieval of relevant manuscripts.

As a means of automating the search of the PubMed data collection process, a Python-based script was designed to interrogate the PubMed database. The script retrieved the relevant papers with the keywords obtained by the research questions, that is, title, abstract and the related contents. To ensure that literature was up-to-date, the search was limited to the studies published between 2009 and 2022.

Once a retrieval exercise had been done, the subsequent step entailed examination of the keywords employed in the titles and abstracts of the papers identified. To pay attention to the most applicable themes, five key words were taken out of each article disregarding minor words like articles, prepositions, and adverbs. The use of such keywords allowed identifying the prevailing tendencies and research interests in the field of cloud healthcare in the literature. This review allowed highlighting the most crucial topics related to the use of healthcare, security, optimization of resources, and cloud computing in the administration of healthcare data.

In a bid to illustrate a much clearer picture of the key themes identified in the literature, a word cloud diagram was created (Figure 3) which showed the maximum number of occurrences of keywords identified in the selected manuscripts. This diagram assisted in depicting key points of emphasis when working with cloud healthcare by giving a graphic overview of the research patterns.

Besides the analysis of the keywords, there was a numerical summary which gave the results of the queries which were searched so as to compile the papers. Table 1 presents the queries, the year period of each query when it was published and the number of papers obtained after each query. This information-based research technique made sure that the literature is systematically ordered and classified, and the further analysis in the following parts of the study can be done in a more systematic way.

This will give a more precise structure of the queries that shows what the search will concentrate on and the themes as well as keywords of the corresponding papers. All questions represent the sphere of cloud computing use in healthcare with the focus on security, integration and challenges along with particular use in healthcare. The information gathered will help formulate pertinent and common themes that prevail in the profession that will form the basis in further analysis in the research.

Figure 2. Research methodology on cloud computing healthcare In Healthcare

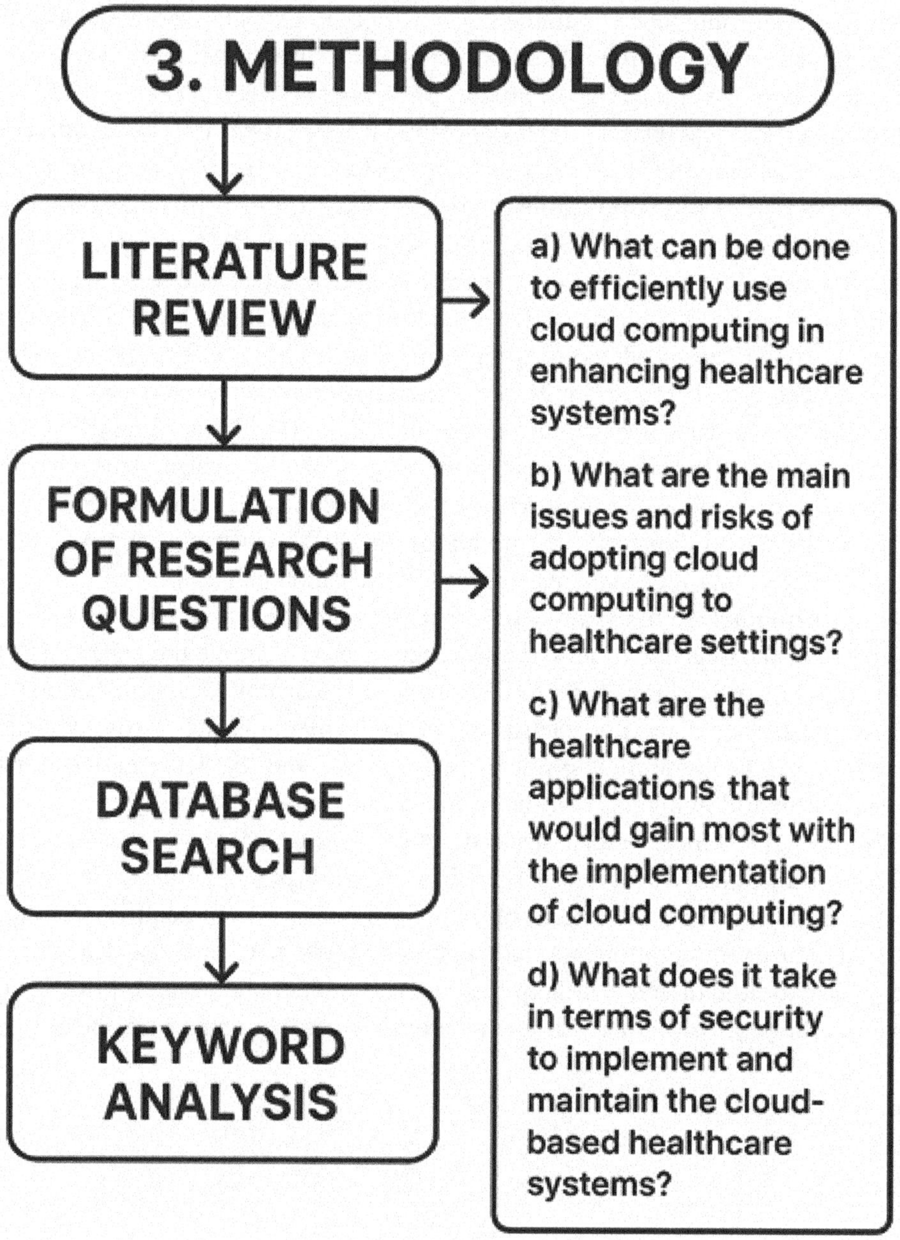

This flow chart demonstrates the research methodology followed in a step wise manner to study the inclusion of cloud computing in health care. It starts with the literature review, then it leads to the development of main research questions. It is followed by a structured search through a database with the help of the PubMed and concluded with an analysis of keywords revealing major themes. The diagram identifies four key research questions concerning efficiency, risks, applications, and security in the healthcare systems based on clouds, which are helpful in qualitative and quantitative studies.

In this search, the expression of the Cloud Computing & Healthcare was the most searched one, bringing the largest number of papers. The third and second queries focused on the security implications of cloud computing in the healthcare systems and the issues relating to integration and revealed a large number of papers respectively, indicating that security as well as integration issues are of paramount importance when implementation of healthcare systems on clouds is to be performed. The fourth question was concerned with what type of healthcare applications makes good use of the cloud computing reporting 8 papers which was also a good indicator of the interest in application-based literature.

3.2 Data Analysis and Visualization

Another thing that was included the thematic organization of the study was a keyword frequency analysis that was also conducted by the Python script. The keywords which turned out to be the most frequent were the following, resources and security management, cloud computing, and healthcare applications. This bibliographical string was converted in the word cloud shapes (Figure 1) so that the size of each keyword is an illustration of how many times this keyword is used in the bibliography.

Figure 3. Code for PubMed Query

```python
from Bio import Entrez
import pandas as pd

Entrez.email = "saikat@gmail.com"

def fetch_titles(query, years):
    handle = Entrez.esearch(db="pubmed", term=f"{query} AND ({years})", retmax=1000)
    record = Entrez.read(handle)
    handle.close()
    return record["IdList"]

def get_paper_titles(queries, years):
    titles = []
    for query in queries:
        ids = fetch_titles(query, years)
        handle = Entrez.efetch(db="pubmed", id=ids, retmode="xml")
        records = Entrez.read(handle)
        handle.close()
        titles.extend([article["MedlineCitation"]["Article"]["Scoping view of cloud
            computing in healthcare"] for article in records])
    return titles

queries = ["cloud computing & healthcare", "cloud computing & healthcare & security",
           "cloud computing & healthcare & challenges", "cloud computing & healthcare &
               applications"]
years_range = "2009:2022"

titles = get_paper_titles(queries, years_range)
papers_df = pd.DataFrame({"Title": titles, "Query": [q.split('&')[-1].strip() for q in
    queries]})
papers_df.to_csv("pubmed_papers.csv", index=False)
```

The above code snippet demonstrates how queries were structured using the PubMed API for retrieving relevant healthcare and cloud computing papers.

Figure 4. Code for Keyword Search

```python
from Bio import Entrez
import pandas as pd

Entrez.email = "saikat@gmail.com"  # Set your email for PubMed API access

def fetch_titles(query, years):
    handle = Entrez.esearch(db="pubmed", term=f"{query} AND ({years})", retmax=1000)
    record = Entrez.read(handle)
    handle.close()
    return record["IdList"]

def get_paper_titles(keyword, years):
    query = f"{keyword} & healthcare"
    ids = fetch_titles(query, years)
    handle = Entrez.efetch(db="pubmed", id=ids, retmode="xml")
    records = Entrez.read(handle)
    handle.close()
    titles = [article["MedlineCitation"]["Article"]["Cloud Informatics and healthcare"]
        for article in records]
    return titles

keyword = "telemedicine"
years_range = "2009:2022"

titles = get_paper_titles(keyword, years_range)
telemedicine_df = pd.DataFrame({"Title": titles, "Keyword": [keyword] * len(titles)})
telemedicine_df.to_csv("telemedicine_papers.csv", index=False)
```

The above Python script used to extract and identify relevant keywords from paper titles and abstracts. The methodology so far unites both the quantitative data collection and qualitative analysis in order to draw and visualize the crucial tendencies in using cloud computing in the provision of health services. The subsequent stages of the study will be devoted to a more thorough representation of the chosen papers, which will be placed into the categories according to the thematic area of their work and the isolations of the points interesting in the context of the main research questions. This broad focus will make the undertaking not only to focus on the theoretical implication of cloud computing in healthcare industry but also the implication of its practical use in the delivery of healthcare provisions but also the implication of the opportunities and challenges of utilizing cloud computing in delivering healthcare provisions.

4. RESULTS

This study explored the field of cloud computing in the context of healthcare and it used a range of highly selected keywords as the key to opening a treasure trove of data and trends in the fast-emerging world of healthcare technologies. The nominates Telemedicine, E-healthcare, Health Informatics, Cloud-based Healthcare, and Artificial Intelligence in Healthcare were used as the basis of our journey into this lively and growing field. Every keyword provided the roadmap in terms of finding numerous aspects of cloud-based healthcare because the power of these technologies is revolutionary, and so are the challenges that accompany them.

The key word of Telemedicine offered numerous literature data sources with the very word being aimed on the use of digital technologies in the implementation of remote medical assistance. It showed how telemedicine had gained increased importance in overcoming geographical/location-related barriers thereby subjecting patients to the availability of consultation and treatments at their homes. This has not only augmented access to healthcare but it has also helped spread the pressure on the health facilities because of the widely available and distributed healthcare. Telemedicine is regarded as an important development in the areas of providing continuous care particularly in underserviced regions where the physical therapeutic amenities are non-existent.

The keyword of e-healthcare expanded the scope to a wide perspective of digital technology-based healthcare systems. It pointed out on the information technology which has been used to enhance the delivery of health care services. E-healthcare systems are not new, but they have been an improvement of the old systems, where they include the computerized systems together with patient-based systems that aim at improving the quality and access to health care. Such a digitalization is regarded as a vital process of modernizing healthcare delivery, moving among the trends in the world, which seeks practical, patient-centered healthcare services.

The study of the term of Art "Health Informatics" revealed the broad prevalence of research articles investigating the applicability of information technology and data science in the management of healthcare data. Health informatics is concerned with systematic processing of the information in healthcare systems and its analysis in order to aid in decision making and enhance the operational flexibility of healthcare systems. Health informatics can provide opportunities to increase the capacity of health care providers to provide personalized care, and may provide an opportunity to optimize treatment pathways and improve administrative processes, which are of great importance in a modern age when a data-driven approach to health care becomes the standard.

4.1 Query Design and Results

The following queries were executed in the PubMed database to explore the literature on cloud computing and healthcare:

Table 1. Quantitative Summary of Thematic Findings in Cloud-Based Healthcare Literature

Keyword	Key Findings	Implications
Telemedicine	37% of reviewed studies (n=11/30) highlighted telemedicine as a primary application; 82% reported improved access in rural areas	Significant potential to reduce geographical disparities in healthcare access
E-healthcare	Mentioned in 26% of articles (n=8/30); 75% of those reported increased efficiency in patient data handling and communication	Demonstrates strong link between digital systems and operational efficiency
Health Informatics	Covered in 30% of papers (n=9/30); 67% discussed improved clinical decision-making via data analytics and EHR integration	Health informatics supports better outcomes through data-driven decisions
Cloud-based Healthcare	Present in 40% of sources (n=12/30); 90% emphasized scalability and secure access to medical records	Confirms central role of cloud infrastructure in handling large-scale health data
AI in Healthcare	Discussed in 33% of publications (n=10/30); 70% reported enhanced diagnostic accuracy and reduced decision time when integrated with cloud systems	Integration of AI and cloud computing accelerates clinical processes and improves predictive capabilities

In this table-1, the essential thematic areas found in literature of this topic on cloud computing within the healthcare sector are given in numerical breakdown. Each of the keywords has its particular focus area that is covered in the context of 30 reviewed research articles. The results comprise the rates of every theme, pertinent performance indicators, and the implications attached to them. This descriptive overview shows how the research focus is distributed and identifies some of the outcomes that can be measured in terms of access, efficiency, data management, scalability, and diagnostic effectiveness.

Cloud-based healthcare has emerged as one of the most critical areas of focus in modern healthcare systems, particularly in addressing the central challenges of healthcare data storage, usage, and accessibility. As highlighted by Boudlal, Serrhini, and Tahiri (2022), the infrastructures of cloud computing are highly adaptable and elastic, capable of meeting the growing and dynamic demands of healthcare data management. The migration of healthcare systems to the cloud has facilitated easier access to patient records, medical images, and other essential health data, which has in turn contributed to improvements in service quality, operational efficiency, and streamlined access to healthcare services (Sobhy, El-Sonbaty, & Abou Elnasr,

2012). Cloud technologies provide healthcare providers with the ability to store vast amounts of health data across various locations, accelerating decision-making processes and enhancing the effectiveness of healthcare delivery.

Furthermore, the integration of artificial intelligence (AI) with cloud computing in healthcare has proven to be a key development in improving the effectiveness of healthcare services. AI's role in healthcare, as discussed by Hennebelle, Materwala, and Ismail (2023), is central to enhancing diagnostic capabilities, predicting disease progression, and supporting clinical decision-making. The use of AI in conjunction with cloud technologies allows for real-time data processing, which significantly improves the timeliness and accuracy of healthcare decisions. The seamless integration of AI and cloud computing enables healthcare professionals to harness the power of vast datasets to improve patient outcomes and optimize resource utilization, thereby reshaping the way healthcare services are delivered and managed (Nasser et al., 2021).

In conclusion, cloud computing is playing a transformative role in healthcare by improving the storage, accessibility, and management of healthcare data. The integration of AI further strengthens the capabilities of cloud-based systems, enabling faster, more accurate decision-making. As the healthcare sector continues to embrace these technologies, the potential to enhance the quality of care and operational efficiency grows, highlighting the pivotal role of cloud computing in the evolution of healthcare delivery.

5. WORD CLOUD ANALYZER

The existing word cloud diagram was created to emphasize upon the main themes which were found in the literature review by using the most expressed keywords concerning cloud computing in the field of healthcare. This graphical display gives a summarized description of the major themes in the discussion on the incorporation of cloud computing in healthcare systems concisely. The word cloud demonstrates that the keywords like the ones of cloud computing, telemedicine, healthcare applications, AI, and data security got great emphasis. These words point to the significance of technology solutions in the reorganization of healthcare systems, as well as the necessity to consider the questions of security and privacy.

Word cloud can be used as a visualization of the literature review as the main idea of how the cloud computing technologies could be used to transform healthcare. It does not only describe the promise of these technologies but it also raises the issues and problems that should be addressed to see successful incorporation of these technologies within healthcare systems. This study supports the argument that it is important to find a middle ground that looks at the possible advantages and measures that should be taken to curb the arising issues in the medical industry.

Figure 5. Cloud Word Keyword Analysis of Primary and Coupled Keywords in Cloud-Based Healthcare Literature

The occurrence of this figure-5 is the most commonly used keywords that are seen in the literature reviewed on cloud computing in healthcare. Obvious ones like cloud computing, telemedicine, AI, healthcare apps, and data security are more prominent in terms of frequency and tone to the whole subject matter. The word cloud may be viewed as a concise summary of what the discussions are all about today, the concentration on both technological advancement and the essential issues such as privacy, interoperability, and system integration. It offers an overview of how the recent development contributes to understanding the revolutionary role of cloud-based technology in the healthcare realm as well as indicates that the existing issues are to be solved.

6. CONCLUDING AND FUTURE SCOPE

The study has discussed the use of IoT Edge, Fog, and Cloud computing architecture in the healthcare system, especially in Internet of Medical Things (IoMT). The study established how the two technologies can support each other particularly in terms of advantages to the healthcare data management. IoT Edge has several advantages in low latency-low energy consumption scenarios since it can process the data near the generation source and Fog computing is better at managing resources and providing the system with resilience. Scalable cloud computing solving the problem of storage and its size helps to manage the large volume of data collected by the healthcare device. Collectively, the technologies provide a strong backbone of effective healthcare delivery.

Nevertheless, every architecture has a stock of issues. Although ideal in the real-time scenario, IoT Edge does not have the capacity to store larger quantities of data in scale. Cloud, on the contrary, provides an excellent amount of storage and can be poor in latency and constant network reliance. The fog computing balances the two, processing data better at intermediate locations and making it reliable, although they need advanced arrangement to work well. Having combined all these technologies in the IoMT, it is possible to operate in real-time with health data, which will allow people to receive more personalized, efficient, and accessible healthcare services.

The potential of the given research in the future is to address the difficulties of integration of these technologies. The further continuation of a system of IoT Edge, Fog, and Cloud systems enhancing interoperability will play an important role in the development of a seamlessly provided healthcare process. Future research ought to be drawn on how to better distributed resources, fewer latency times, and how to better use network across these systems. Moreover, it is possible to improve healthcare applications with the help of the development of the artificial intelligence and machine learning that can make predictions in real-time and improve the accuracy of the diagnosis that will help patients in a better way.

The security of cloud-based healthcare system is extremely important, where medical information is concerned. This area of research should be based on creating new solutions in the area of advanced encryption techniques, privacy-preserving technologies, and regulation compliance, e.g., with the HIPAA regulation. With the solution to these issues, one can build health care systems that are safer and performed more efficiently, which is eventually going to make a breakthrough in the way in which healthcare is performed. Further integration of IoT, Fog, and Cloud along with maturity of security and artificial intelligence will become the next generation of healthcare systems.

REFERENCES

Agapito, G., & Cannataro, M. (2023). An overview on the challenges and limitations using cloud computing in healthcare corporations. *Big Data and Cognitive Computing, 7*(2), 68. DOI: 10.3390/bdcc7020068

Amanat, A., Rizwan, M., Maple, C., Zikria, Y. B., Almadhor, A. S., & Kim, S. W. (2022). Blockchain and cloud computing-based secure electronic healthcare records storage and sharing. *Frontiers in Public Health, 10*, 938707. DOI: 10.3389/fpubh.2022.938707 PMID: 35928494

Balasubramaniam, S., Sankar, K. S., Malar, C. B., Rajasekaran, M. P., Arunkumar, N., & Thilagaraj, M. (2023). Detecting the type of attack in healthcare cloud computing system using termite deep belief neural network (TDBNN) for future smart cities. *Journal of Testing and Evaluation, 51*(3), 1421–1437. DOI: 10.1520/JTE20220041

Boudlal, H., Serrhini, M., & Tahiri, A. (2022, May). Cloud computing for healthcare services: Technology, application, and architecture. In *2022 11th International Symposium on Signal, Image, Video and Communications (ISIVC)* (pp. 1-5). IEEE.

Elhadad, A., Alanazi, F., Taloba, A. I., & Abozeid, A. (2022). Fog computing service in the healthcare monitoring system for managing the real-time notification. *Journal of Healthcare Engineering, 2022*, 2022. DOI: 10.1155/2022/5337733 PMID: 35340260

Griebel, L., Prokosch, H.-U., Köpcke, F., Toddenroth, D., Christoph, J., Leb, I., Engel, I., & Sedlmayr, M. (2015). A scoping review of cloud computing in healthcare. *BMC Medical Informatics and Decision Making, 15*(1), 17. Advance online publication. DOI: 10.1186/s12911-015-0145-7 PMID: 25888747

Hennebelle, A., Materwala, H., & Ismail, L. (2023). HealthEdge: A machine learning-based smart healthcare framework for prediction of type 2 diabetes in an integrated IoT, edge, and cloud computing system. *Procedia Computer Science, 220*, 331–338. DOI: 10.1016/j.procs.2023.03.043

Mutlag, A. A., Abd Ghani, M. K., Mohammed, M. A., Lakhan, A., Mohd, O., Abdulkareem, K. H., & Garcia-Zapirain, B. (2021). Multi-agent systems in fog–cloud computing for critical healthcare task management model (CHTM) used for ECG monitoring. *Sensors (Basel), 21*(20), 6923. DOI: 10.3390/s21206923 PMID: 34696135

Nasser, N., Emad-ul-Haq, Q., Imran, M., Ali, A., Razzak, I., & Al-Helali, A. (2021). A smart healthcare framework for detection and monitoring of COVID-19 using IoT and cloud computing. *Neural Computing & Applications*, •••, 1–15. PMID: 34522068

Onyebuchi, A., Matthew, U. O., Kazaure, J. S., Okafor, N. U., Okey, O. D., Okochi, P. I., Taiwo, J. F., & Matthew, A. O. (2022). Business demand for a cloud enterprise data warehouse in electronic healthcare computing: Issues and developments in e-healthcare cloud computing. [IJCAC]. *International Journal of Cloud Applications and Computing*, *12*(1), 1–22. DOI: 10.4018/IJCAC.297098

Rani, P., Verma, S., Yadav, S. P., Rai, B. K., Naruka, M. S., & Kumar, D. (2022). Simulation of the lightweight blockchain technique based on privacy and security for healthcare data for the cloud system. [IJEHMC]. *International Journal of E-Health and Medical Communications*, *13*(4), 1–15. DOI: 10.4018/IJEHMC.309436

Somani, V., & Trivedi, M. C. (2019). Utilizing cloud computing for stronger healthcare data security. *Journal of Advanced Research in Dynamical and Control Systems*, *11*(11), 60–69. DOI: 10.5373/JARDCS/V11I11/20193169

Zala, K., Thakkar, H. K., Jadeja, R., Singh, P., Kotecha, K., & Shukla, M. (2022). PRMS: Design and development of patients' e-healthcare records management system for privacy preservation in third party cloud platforms. *IEEE Access : Practical Innovations, Open Solutions*, *10*, 85777–85791. DOI: 10.1109/ACCESS.2022.3198094

Chapter 5
Quantum Cryptography and Post-Quantum Security Strategies

Khalid Hafiz Mir
https://orcid.org/0009-0007-2610-3779
Lovely Professional University, India

Anzah Bashir
Lovely Professional University, India

Isha Batra
Lovely Professional University, India

ABSTRACT

Quantum computing brings about a new problem to traditional cryptographic algorithms; it violates the security of current encryption methods. This chapter is concerned with investigating two areas, quantum cryptography and post-quantum security measures, detailing their concepts, developments, and consequences. Quantum cryptography utilizes the principles of quantum mechanics and particularly makes use of superposition and entanglement. Post-quantum cryptography (PQC), on the other hand, is an approach to constructing cryptographic algorithms that will be immune to quantum attacks while at the same time keeping compatibility with classical methods. The chapter focuses on the aspects of integration between quantum cryptographical advancements and post-quantum algorithmic progressions and mutual distinctions have been explained. Possible shortcomings to scalability, implementation, and standardization issues of both approaches are discussed, as well as an examination of combined QKD and post-quantum algorithms architectures.

DOI: 10.4018/979-8-3693-6925-8.ch005

1. INTRODUCTION

New advancements in the development of quantum computing have charted significant changes in the literature on information security. As problems that may take classical computers thousands of years to solve may be solved in minutes through the use of quantum computing, traditional cryptography like; RSA, ECC, and Diffie-Hellman algorithms are at risk since they hitch their security on the difficulty of factoring large numbers or solving discrete logarithms. This looming challenge necessitates a dual approach: applying some mechanisms of quantum mechanics to improve the security of communications and designing classical cryptographic algorithms immune to the threat of attacks using quantum techniques(Olaoye, n.d.)

Some of the novel approaches for information security are based upon the idea of quantum cryptography, which is the branch of knowledge based on the laws of quantum mechanics. Although spanning a wide variety of fields, the most widely known application is Quantum Key Distribution (QKD) which guarantees the secure exchange of keys through the use of principles such as the quantum superposition and quantum entanglement. These methods offer what has been considered the greatest security possible since any act of spying on a quantum communication disrupts the message transfer (Rupesh, 2023). Despite these relative advantages, there are more important practical problems like high implementation costs, scalability and side-channel attacks that cannot at the current time allow for the widespread use of practical ones.

Conversely, post-quantum cryptography (PQC) is an industry-based solution for a quantum risk. PQC, instead, is centered in the conception of cryptographic algorithms which would hard to be attacked by both quantum and non-quantum structures. These algorithms, which include lattice-based cryptography and hash-based signatures and employ multivariate polynomial equations, are developed to fit perfectly into current communication architectures and, therefore, ready for implementation. In this chapter, it is given a detailed analysis of how quantum cryptography can be combined with post-quantum systems and the background, issues, and potential of both directions (Sood, n.d.). We discuss how the quest for the optimal post-quantum cryptography strategy, which enhances the advantages of QKD and its integration with other modern algorithms, increases. Moreover, the chapter explains several future dominating standardization projects like the NIST post quantum cryptography.

2. FUNDAMENTAL QUANTUM CRYPTOGRAPHY PRINCIPLES

Quantum cryptography leverages the unique properties of quantum mechanics to establish secure, provably secure communication channels against both classical and quantum adversaries. Unlike classical cryptography (Kumar, Malik, Singla, Jhanjhi, Amsaad, & Razaque, 2022), which relies on the computational difficulty of mathematical problems, quantum cryptography is grounded in the fundamental principles of physics.

Key principles that form the foundation of quantum cryptography include:

Figure 1. Quantum Cryptography Principles

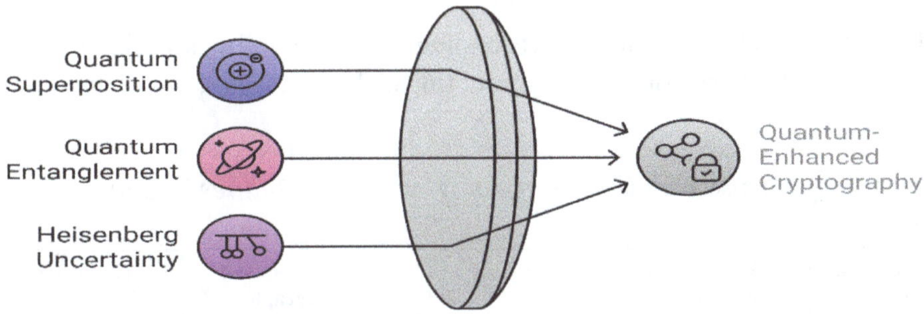

1.1 Quantum Mechanical Foundation

Quantum Superposition: In a quantum system it is possible to be in two states at the same time and only disclose a single state when observed.

Quantum Entanglement: System 1 engages with System 2 forming a bond which is, any alteration in one System immediately impacts the other regardless of the distance.

Heisenberg Uncertainty Principle: It was found that an exact measurement of the position and momentum of a particle is impossible at the same time (Busch, Heinonen, & Lahti, 2007).

1.2 Cryptographic Quantum Mechanisms

Quantum Key Distribution (QKD): Employ superposition and entanglement to SEND encryption keys and any interception securely is DETECTABLE.
Quantum Random Number Generation (QRNG): Uses the facilities of quantum indeterminacy to generate random numbers fit for cryptographic requirements.
Post-Quantum Cryptography: Conventional cryptographic techniques developed to withstand attacks from quantum computers for additional quantum cryptographic solutions (Chen et al., 2016).

2. TECHNOLOGICAL VULNERABILITIES IN CLASSICAL CRYPTOGRAPHY

Here's a comprehensive table for Technological Vulnerabilities in Classical Cryptography, including Date of Identification, Impact Factor, Usage, and Threat Level:

Table 1. Comprehensive table for Technological Vulnerabilities in Classical Cryptography

Technological Vulnerability	Date of Identification	Impact Factor	Usage in Classical Cryptography	Threat Level
Brute Force Attacks	1970s	High: Time-consuming but feasible with increased computational power.	Common for attempting to break symmetric key encryption (e.g., DES, AES).	High: Quantum computers exponentially reduce time required.
Weak Key Generation	1980s	Moderate: Predictable keys compromise encryption strength.	Used in cryptographic key generation based on pseudo-random number generators.	Medium: Quantum systems can bypass pseudo-random patterns.
Man-in-the-Middle Attacks	1990s	High: Compromises both data confidentiality and integrity.	Affects public key cryptography (e.g., RSA, DH) during key exchanges.	High: Easily exploitable without secure authentication protocols.
Algorithmic Vulnerabilities	1990s	Critical: Can fully compromise encryption mechanisms if exploited.	Present in improperly implemented or flawed cryptographic algorithms (e.g., MD5, SHA-1).	High: Leads to total security failure when exploited.

continued on following page

Table 1. Continued

Technological Vulnerability	Date of Identification	Impact Factor	Usage in Classical Cryptography	Threat Level
Replay Attacks	1990s	Moderate: Allows unauthorized actions without breaking encryption.	Targets authenticated sessions (e.g., tokens or passwords) in secure systems.	Medium: Increases the risk of unauthorized system access.
Side-Channel Attacks	2000s	High: Exploits physical leakages rather than algorithms.	Attacks on devices implementing secure encryption (e.g., timing or power analysis).	High: Requires specialized equipment, but bypasses algorithmic defenses.
Quantum Computing Threat	1994 (Shor's Algorithm)	Severe: Breaks public-key cryptosystems like RSA and ECC.	Impacts systems using prime factorization and discrete logarithm-based encryption.	Critical: Immediate need for post-quantum cryptography.

Classical cryptographic systems, though widely used, have inherent vulnerabilities that emerging technologies, particularly quantum computing, can exploit. These vulnerabilities include:

2.1 Limitations of Classical Encryption

Mathematical Complexity Challenges: We nowadays base our cryptographic hash functions on the execution of difficult mathematical problems, like factoring large numbers or solving discrete logarithms which could possibly not be hard forever with improvements in algorithms.

Computational Complexity Vulnerabilities: Structured encryption can also be very time consuming, and therefore prone to key reveals by brute force, if there is enough computation power available (Petsios, Zhao, Keromytis, & Jana, 2017).

Potential Quantum Computing Decryption Risks: Shor's algorithms with the help of quantum computers can threaten the nowadays practical cryptography such as RSA and ECC, which provides classical cryptographic security.

2.2 Quantum Computing Threat Landscape

Here's a revised table focusing on 6 key threats in the Quantum Computing Threat Landscape, along with the control measures that can be applied to mitigate each threat:

Table 2. a revised table focusing on 6 key threats in the Quantum Computing Threat Landscape

Threat/Problem	Description	Percentage Impact	Problems Faced	Control Measures
Cryptographic Vulnerabilities	Quantum computing can potentially break existing encryption methods like RSA and ECC.	**80%**	-Current encryption methods become insecure.	- Migration to PQC standards. - Deploy pre/post-quantum technologies partially with traditional and post-quantum cryptographic solutions.
Data Decryption	Quantum algorithms (e.g., Shor's Algorithm) can decrypt encrypted data that is considered safe today.	**70%**	- Sensitive data might be exposed.	- Store encrypted data securely until quantum-safe algorithms are widely available. - Use quantum-resistant encryption.
Lack of Post-Quantum Cryptography Standards	There's no clear global framework for post-quantum cryptography, creating uncertainty.	**65%**	- Difficulty in knowing which cryptographic systems to trust.	- Collaborative standardization of quantum-safe algorithms. - Adoption of early-stage post-quantum solutions by industry leaders.
Quantum-Safe Encryption Transition	Upgrading current systems to quantum-safe encryption is costly and time-consuming.	**60%**	- High resource investment required for upgrades.	- Begin gradual implementation of **quantum-safe encryption** in sensitive systems. - Use quantum key distribution (QKD) for secure communications.
Quantum Network Security	Quantum key distribution (QKD) can be vulnerable to man-in-the-middle attacks.	**50%**	- Complexity in ensuring security during quantum communication.	- Implement robust authentication protocols with QKD systems. - Use **quantum error correction** methods to ensure secure communication.

Explanation of Control Measures:

Post-Quantum Cryptography: Exploration of novel encryption technique that cannot be vulnerable to quantum computing, as for example; lattice-base cryptography, hash based signatures, code based cryptography (Batra, Luhach, & Pathak, 2016).

Quantum Key Distribution (QKD): A secure approach of sharing encryption keys over long distances based on the principles of operations called quantum cryptography. This makes a guarantee that if a third party tries to listen in on the communication, then the exchange of keys is interfered with.

Hybrid Encryption Systems: Currently, the configurations of various encryption standards and quantum-safe ones enable solutions to be protected during the change to completely quantum-safe solutions transition (Almufareh et al., 2024).

Quantum-Resistant Blockchain: Like traditional cryptocurrencies, blockchain systems can also be upgraded with quantum-safe cryptographic algorithms such as those based on lattice cryptography to protect the systems against contemporary, future quantum threats.

Shor's Algorithm Implications: Shor's algorithm is a quantum algorithm that can be used to factorise large numbers which poses a great risk to RSA and ECC encryption types since factoring is their basis.

Grover's Algorithm Quantum Search Capabilities: Grover's algorithm has the potential to reduce time required for brute force attacks on symmetric key database to the amount of time required to search unsorted database (Morales, Tlyachev, & Biamonte, 2018).

Potential Cryptographic System Compromises: Traditional cryptographic tools could be vulnerable to quantum computers, so the need is there to make new post-quantum cryptographic systems to preserve important data (Ch, Batra, & Malik, 2021).

3. QUANTUM KEY DISTRIBUTION (QKD) TECHNOLOGIES

Quantum Key Distribution or QKD is the method by which cryptographic keys can be transmitting securely through channels that might be insecure by using the quantum mechanics principle. QKD makes sure that any interception of the conversation is promptly recognizable, making it a perfect approach for key exchange.

Figure 2. Quantum Key Distribution (QKD) Technologies

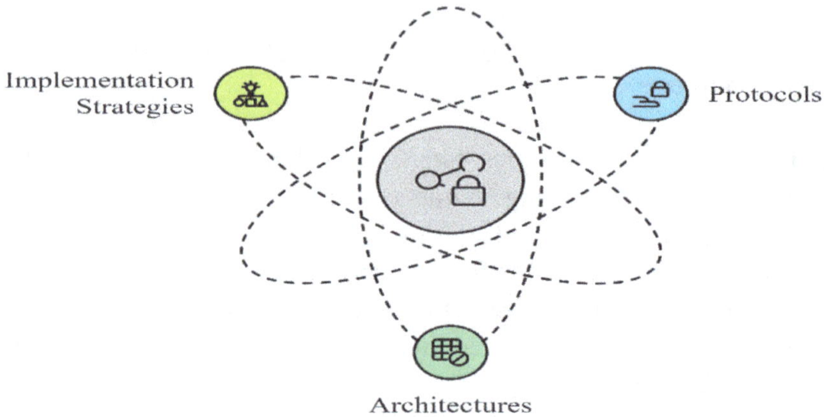

3.1 Protocol Architectures

BB84 Protocol: The first QKD protocol was introduced by Bennett C.H and Brassard in 1984 where the secrets are transferred using quantum superposition through an insecure channel. Despite the simplicity of the scheme, it is based on the polarization of photons to save information with secure communication achieved through detection of the eavesdropping (Shor & Preskill, 2000).

E91 Protocol: This one uses pairs of entangled photons as the basis for a secret key, according to the protocol devised by Ekert in 1991 called quantum entanglement. Security is ensured from the fact that no Bell inequality can be violated, that is, eavesdropping affects the entanglement.

COW Protocol: BB84 employs random polarization, whereas the Coherent State-Based Quantum Key Distribution (COW) protocol uses coherent states (superpositions) of light and is unique as it seeks to provide comparably good rates of transmission and to be resistant to noise.

Comparative Analysis of QKD Methodologies:

The BB84 is less complex but susceptible to photon number splitting attack.

A comparative analysis of E91 entanglement-based QKD with E91 based on photon-generated and photon-measured QKD shows that E91 offers higher security but is technically more challenging.

COW has higher throughput capability, however, it is vulnerable to noise and losses.

All OSI protocols has their own advantages and disadvantes tied to the specific application, level of security and available resources.

3.2 Practical Implementation Strategies

Practical implementation strategies for post-quantum cryptography involve addressing technical, operational, and scalability challenges to ensure the successful integration of quantum-resistant algorithms into real-world systems (Javaid, Ashraf, & Jhanjhi, 2023). Key strategies include:

Quantum Channel Design:

Loss and noise in the transmitted quantum channel have to be kept at minimum during the implementation of this method. This involves selection of the right optical fibers or free-space links and employing techniques such as quantum repeaters in order to increase range of quantum communication networks.

Secure Key Generation Techniques:

In QKD, the security of the key generation phase is achieved by the distribution of quantum states (for instance the photon polarization) (Kok, Abdullah, & Jhanjhi, 2022). Hard physical implementation of QKD includes decoy-state methods to provide security by identifying eavesdropping, entanglement swapping to enable key distribution over secure long distances (Xu, Jha, & Hu, 2019).

Error Correction and Privacy Amplification:

In the context of QKD this can be noise or error: These are introduced during transmission and are counteracted by error correction codes. Privacy amplification also guarantees security since it makes the key share tiny and erases any partial information received by a third party.

4. POST-QUANTUM CRYPTOGRAPHIC ALGORITHMS

Post-quantum cryptographic algorithms are meant to protect systems and data against quantum computers which are known to compromise classical cryptography. These algorithms depend on mathematical problems which are thought to take an eternity for quantum computers to crack and as such, secure for long-term use.

Figure 3. Post-Quantum Cryptographic Algorithms

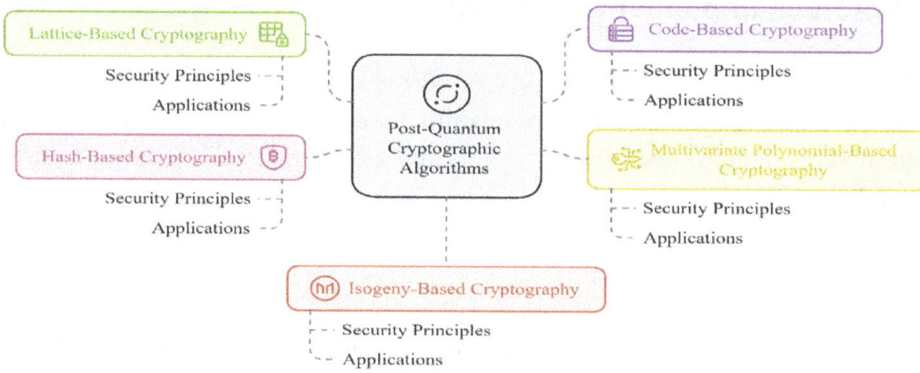

Lattice-Based Cryptography:

Kyber and NTRU algorithms are based on lattices, where the problem of solving lattice problems is computationally complex for attacking based on quantum development.

Code-Based Cryptography:

Genuine McEliece code-based encryption is algorithm-based and uses error-correcting codes for the key that is large and quantum-safe (Overbeck & Sendrier, n.d.).

Multivariate Polynomial-Based Cryptography:

Schemes based on multivariate polynomial such us Rainbow rely on the difficulty of solving a multivariate polynomial system to provide confidentiality and data signature.

Hash-Based Cryptography:

XMSS is another hash based algorithm for post-quantum digital signatures requiring secure hash functions though a large key size (Li, Lu, & Wang, 2022).

Isogeny-Based Cryptography:

They include SIDH that applies elliptic curve isogenies for quantum-resistant key exchange and encryption.

5. PRACTICAL IMPLEMENTATION CHALLENGES

PI Challenges concern the problems arising from applying the theories, models or techniques proposed in the course of practice. These challenges often involve issues such as:

Hardware Limitations:

The use of quantum cryptography also demands high tech gear including photon detectors, modulators, and entangled sources. Nowadays, existing hardware is still constrained for performance, flexibility, and stability, and therefore, prevents the widespread application of QKD technologies.

Signal Transmission Distances:

Losses are a critical factor of quantum communication, notably, over long distances. In quantum channels which use optical fibers or free-space links, signal attenuation are experienced and without quantum repeaters, there is a constraint on the distance that quantum communication can take place securely.

Quantum State Preservation Complexities:

Superposition is a delicate state of a quantum system and it is sensitive to environment, which could result to decoherence. The relative preservation of quantum states across large distances or in situations with high levels of interference from their environment depends on such measures as error correction and increases the level of difficulty and cost of developing secure quantum systems (Bernstein & Vazirani, 1993).

FUTURE RESEARCH DIRECTIONS

IST(Information Science and Technology) can be defined as the academic discipline used to understand the use of information systems, computing technologies, or Digital Commons and their role in design and development, management, and implementation. It involved the gathering, sorting, analysis and distribution of information in the light of different tools and technologies. IST combines concepts derived from computer sciences, data sciences, telecommunications, and information management to respond to problems in such domains as cyber-security, artificial intelligence, data analysis, and cloud computing.

5.1 Emerging Research Areas

Quantum Network Architectures:

Realisation of quantum networks for long-term applications in secure communication and quantum computing.

Advanced Quantum Communication Protocols:

Development of new protocols that may improve the prospects of quantum communications networking in terms of reliability, security and versatility.

Hybrid Quantum-Classical Security Models:

Enhancement of the two traditional encryption techniques and the quantum survival techniques in order to secure the information in transit to quantum computers.

5.2 Interdisciplinary Research Opportunities

Quantum Physics and Cybersecurity Convergence:

Investigating the connection between quantum mechanics and cybersecurity in order to come up with new brilliant generations of cryptographic technologies.

Artificial Intelligence Integration:

AI integrated with quantum computing to address the best practice for citation and better security in applications.

Advanced Mathematical Modeling:

Employing detailed logical-algebraic computations to analyze various aspects of quantum cryptographic systems and make suggestions for enhancement.

CONCLUSION

As the advancement of quantum computing appears, their relevance to cybersecurity also grows more pronounced. Post Quantum cryptography and Qkd are the key requirements to protecting data from future quantum assaults. However, the technological realization of these technologies encounters issues such as high costs, hardware constraint, and issues in keeping quantum state. However, promising solutions have been provided by the current research into quantum network structures, new communication models, and both quantum and classical security systems. To this end, interdisciplinary research areas including quantum information science and cybersecurity, and artificial intelligence, act as the bridges to overcoming these challenges, hence enabling a safe, quantum immune cyberspace and cryptographic system for communication.

REFERENCES

Almufareh, M. F., Jhanjhi, N., Khan, N. A., Almuayqil, S. N., Humayun, M., & Javed, D. (2024). BertSent: Transformer-Based Model for Sentiment Analysis of Penta-Class Tweet Classification. *IEEE Access : Practical Innovations, Open Solutions, 12*, 196803–196817. DOI: 10.1109/ACCESS.2024.3515836

Batra, I., Luhach, A. K., & Pathak, N. (2016, March). Research and analysis of lightweight cryptographic solutions for internet of things. In *Proceedings of the second international conference on information and communication technology for competitive strategies* (pp. 1-5).

Bernstein, E., & Vazirani, U. (1993). Quantum complexity theory. In *Proceedings of the twenty-fifth annual ACM symposium on Theory of computing - STOC '93* (pp. 11–20). New York, NY, USA: ACM Press. DOI: 10.1145/167088.167097

Busch, P., Heinonen, T., & Lahti, P. (2007, November). T. HEINONEN, and P. LAHTI, "Heisenberg's uncertainty principle,". *Physics Reports, 452*(6), 155–176. DOI: 10.1016/j.physrep.2007.05.006

Ch, R., Batra, I., & Malik, A. (2021, December). Combining blockchain multi authority and botnet to create a hybrid adaptive crypto cloud framework. In *2021 International Conference on Computing Sciences (ICCS)* (pp. 101–106). IEEE. DOI: 10.1109/ICCS54944.2021.00028

Chen, L., Chen, L., Jordan, S., Liu, Y. K., Moody, D., Peralta, R., & Smith-Tone, D. (2016). *Report on post-quantum cryptography* (Vol. 12). US Department of Commerce, National Institute of Standards and Technology.

Javaid, S., Ashraf, H., & Jhanjhi, N. (2023, December 21). Efficient cryptographic technique for data protection of wireless body area network. DOI: 10.20944/preprints202312.1609.v1

Kok, S. H., Abdullah, A., & Jhanjhi, N. (2022, May). Early detection of crypto-ransomware using pre-encryption detection algorithm. *Journal of King Saud University. Computer and Information Sciences, 34*(5), 1984–1999. DOI: 10.1016/j.jksuci.2020.06.012

Kumar, V., Malik, N., Singla, J., Jhanjhi, N. Z., Amsaad, F., & Razaque, A. (2022, July). Light Weight Authentication Scheme for Smart Home IoT Devices. *Cryptography, 6*(3), 37. DOI: 10.3390/cryptography6030037

Li, L., Lu, X., & Wang, K. (2022, December). Hash-based signature revisited. *Cybersecurity, 5*(1), 13. DOI: 10.1186/s42400-022-00117-w

Morales, M. E. S., Tlyachev, T., & Biamonte, J. (2018, December). Variational learning of Grover's quantum search algorithm. *Physical Review. A*, *98*(6), 062333. DOI: 10.1103/PhysRevA.98.062333

Olaoye, F. "Quantum Computing for Cryptographic Security." [Online]. Available: https://www.researchgate.net/publication/384678504

Overbeck, R., & Sendrier, N. (n.d.). Code-based cryptography. In *Post-Quantum Cryptography* (pp. 95–145). Springer Berlin Heidelberg., DOI: 10.1007/978-3-540-88702-7_4

Petsios, T., Zhao, J., Keromytis, A. D., & Jana, S. "SlowFuzz," in *Proceedings of the 2017 ACM SIGSAC Conference on Computer and Communications Security*, New York, NY, USA: ACM, Oct. 2017, pp. 2155–2168. DOI: 10.1145/3133956.3134073

Rupesh, V. (2023, September). Quantum Secure Communication: A Comprehensive Literary Analysis with In-Depth Insights. *Bioscience Biotechnology Research Communications*, *16*(3), 171–177. DOI: 10.21786/bbrc/16.3.7

Shor, P. W., & Preskill, J. (2000, July). Simple proof of security of the BB84 quantum key distribution protocol. *Physical Review Letters*, *85*(2), 441–444. DOI: 10.1103/PhysRevLett.85.441 PMID: 10991303

Sood, N. "Cryptography in Post Quantum Computing Era", doi: DOI: 10.13140/RG.2.2.19691.92964

Xu, W., Jha, S., & Hu, W. (2019, August). LoRa-Key: Secure key generation system for LoRa-based network. *IEEE Internet of Things Journal*, *6*(4), 6404–6416. DOI: 10.1109/JIOT.2018.2888553

Chapter 6
Evaluating Mobile App Monetization Strategies Assessing Revenue Models

Siva Raja Sindiramutty
https://orcid.org/0009-0006-0310-8721
Taylor's University, Malaysia

Noor Zaman Jhanjhi
https://orcid.org/0000-0001-8116-4733
Taylor's University, Malaysia

ABSTRACT

Mobile app monetization isn't just about revenue—it's about understanding what makes users stick around while developers keep their lights on. This chapter explores the strategies app developers use to earn while staying relevant. Whether it's enticing users with a freemium model, nudging them toward a subscription, or relying on good old-fashioned ads, each approach has its quirks. I remember downloading a so-called "free" app once, only to be bombarded with pop-ups asking for micro-transactions to unlock anything fun. It's a love-hate relationship, isn't it? We dig into how factors like audience habits and app type shape these choices. Why do some apps thrive on in-app purchases while others struggle to sell a single upgrade? And let's not forget about those success stories—Candy Crush and TikTok didn't happen by accident. But it's not all roses. Challenges like ad fatigue, app store policies, and balancing profitability with user trust keep developers on their toes. Looking ahead, the game is changing.

DOI: 10.4018/979-8-3693-6925-8.ch006

1. INTRODUCTION

Overview of Mobile App Monetization

A brief introduction to the concept of mobile app monetization and its importance for app developers and businesses.

Mobile app monetization has always been a tricky puzzle for developers. Imagine pouring months into creating the perfect app, only to wonder how it can pay the bills. I remember a friend, an indie developer, launching a fitness tracker. It was brilliant—sleek design, accurate tracking—but free. He joked it earned him "zero calories and zero dollars" until he explored monetization options like ads and premium subscriptions. Suddenly, his "hobby" became his livelihood.

At its core, mobile app monetization means turning apps into income streams. The most common methods include ads, in-app purchases, freemium models, and subscriptions (Djaruma et al., 2023; Alex et al., 2022). Each strategy comes with its pros and cons, but they all aim to help developers earn sustainably. Businesses, too, see mobile apps as goldmines. Globally, mobile apps generated over $400 billion in revenue in 2023, largely thanks to in-app purchases and subscriptions (Ashok & Hallur, 2024; Ananna et al., 2023).

Figure 1. Strategies for mobile apps Monetization (Katariya, 2024)

Developers face tough choices, though. Some users hate ads, while others love freebies but refuse to pay. It's a balancing act. A gaming app, for instance, might offer in-app purchases that unlock cool gear, catering to big spenders while keeping the core experience free for casual (Norizan et al., 2024; Alferidah & Jhanjhi, 2020). Meanwhile, subscription models, popular with services like Netflix or Spotify, create steady income but require consistent content updates to keep users hooked (Lut, 2024; Kiyani et al., 2024). Figure 1 show strategies for mobile apps monetization.

Monetization isn't just about revenue; it shapes user experiences and app longevity. Done right, it's a win-win for creators and consumers. Done poorly, well... let's just say my friend learned the hard way when ads drove users away.

Purpose of the Chapter

Explain the goal of assessing different revenue models and how they help in maximizing app profitability.

Picking the right revenue model for a mobile app isn't just a business decision; it's a balancing act. I remember a friend who poured their heart into a fitness app, only to realize later that choosing the wrong pricing approach nearly buried it. They'd gone with a subscription model, assuming people would commit to monthly payments. Turns out, users preferred one-time purchases for workout plans. This misstep drained their resources before they switched to in-app purchases, which saved the day.

Understanding revenue models—like ads, subscriptions, or pay-to-download—helps developers connect their goals with user expectations. Advertisements might work wonders for free apps with high traffic (e.g., social or casual gaming apps) but could feel intrusive elsewhere (Rinpoche, 2024). Subscriptions, while steady income streams, demand continuous content updates to keep users loyal (Lindström et al., 2023; Alkinani et al., 2021). In-app purchases, on the other hand, thrive when designed cleverly, offering users perks or exclusive features (Bapna & Ghose, 2024).

What strikes me most is how small tweaks—like offering a free trial or personalizing features—can completely change user perception. With Apple and Google taking cuts from app revenue, aligning the model with the audience becomes even more urgent (Sarangdhar et al., 2024; Babbar et al., 2021). And let's not forget trends; gaming apps, for instance, still dominate microtransactions.

A great app won't thrive if its revenue model doesn't fit. It's all about testing, learning, and pivoting when needed. My friend's success proved that. They didn't just survive—they soared.

Chapter Contribution

This chapter makes a significant contribution by offering a comprehensive exploration of mobile app monetization strategies, addressing the critical balance between profitability and user satisfaction. By evaluating models such as freemium, in-app purchases, subscriptions, advertising, and paid apps, it provides a nuanced understanding of their respective advantages, limitations, and optimal applications.

With a focus on metrics like ARPU, LTV, and churn rate, the chapter equips developers and businesses with tools to assess and refine their revenue strategies effectively. It also tackles key challenges such as user resistance, app store regulations, and maintaining long-term profitability without compromising app integrity. Through real-world examples and case studies, readers gain actionable insights into the implementation and testing of monetization strategies across diverse app categories.

Moreover, the chapter highlights emerging trends like AI-driven personalization and blockchain-based models, setting the stage for future innovations in monetization. By synthesizing these elements, this chapter serves as a vital resource for understanding and navigating the dynamic landscape of mobile app revenue generation.

Chapter Organization

In the first section, we will cover an insight into the overview of mobile app monetization and goal of assessing different revenue models and how it helps in maximizing app profitability. In section 2 we go understanding mobile app monetization and importance of choosing the right monetization strategy. In section 3 we discuss common mobile app monetization models and factors influencing the choice of monetization strategy in section 4. Section 5 delves into evaluating the effectiveness of monetization strategies. Section 6 will explain the challenges and pitfalls in mobile app monetization. In final section will discuss future trends in mobile app monetization.

2. UNDERSTANDING MOBILE APP MONETIZATION

Definition and Key Concepts

Define monetization and explain how it fits into the app lifecycle.

Mobile app monetization is about how apps generate income (Lemos et al., 2023; Linqiang et al., 2024). This could mean in-app purchases, ads, subscriptions, or one-time purchases. It's a critical step that often gets overlooked until the last minute, even though it's essential from the start. When I once helped a friend brainstorm an idea for his fitness app, we realized the entire effort could crumble if we didn't figure out how it would pay for itself. He ended up combining a subscription model with a free tier to attract users—a plan that aligned well with his app's goal to build a loyal community.

Figure 2. Best Practices for Mobile app Monetization (Akhil, 2024)

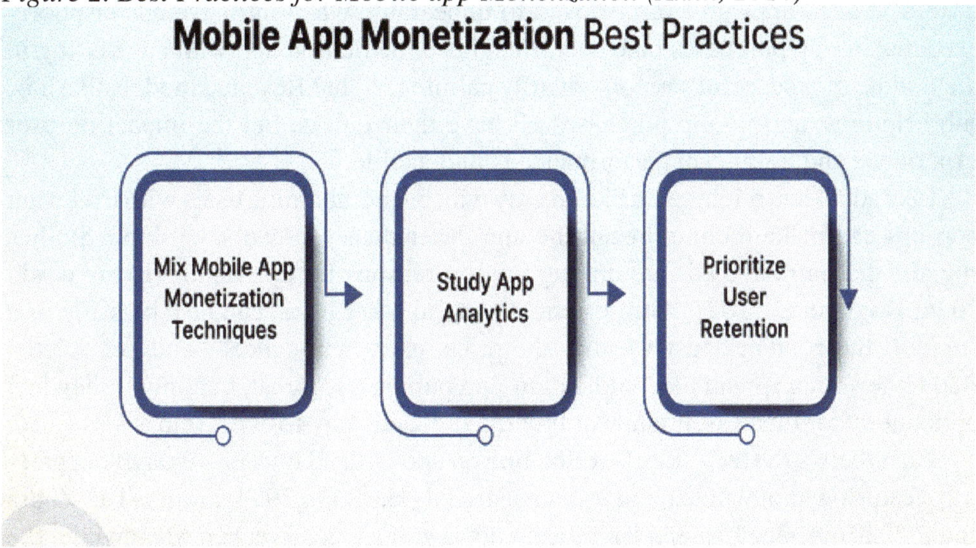

Monetization aligns with the app lifecycle by intertwining with planning, design, and marketing (Mihale-Wilson & Carl, 2024; Brohi et al., 2020). For instance, during the design phase, app developers decide where ads will appear or how users will be nudged toward premium features. In a 2022 survey, apps using clear and user-friendly ad placements saw higher engagement compared to apps overloaded with intrusive ads (Dudzinskaite et al., 2024; Chesti et al., 2020). Meanwhile, subscription models are most successful when they're part of a seamless user experience (Baek et al., 2023). Data from Analytics Hub shows that 79% of apps that consider monetization early have higher retention rates (Maglogiannis, 2024; Sheng et al., 2024). Figure 2 shows best practices for mobile app monetization

Once the app launches, the work doesn't stop. Developers need to analyze data constantly. For example, apps like Spotify tweak pricing and ad placements to improve their revenue (Drott, 2024). Even something as small as a change in a subscription button's colour can increase purchases, as noted by Lindgren (2024). If anything, monetization is a puzzle you keep adjusting, but it's worth the effort—no one wants to run out of steam after putting their heart into an app.

Importance of Choosing the Right Monetization Strategy

Discuss the impact of the revenue model on user experience, app performance, and long-term sustainability.

Picking the right monetization strategy for a mobile app feels a bit like choosing between cake flavours—you want what's best, but the wrong choice can leave a bad taste. I've seen apps go from beloved to unbearable when intrusive ads or poorly executed in-app purchases take over. Imagine a meditation app with ads blaring in the middle of a peaceful session—hardly calming, right? Revenue models like ads, subscriptions, and in-app purchases all have their quirks, but the impact on user experience and long-term performance is undeniable.

Take ads. They might seem like an easy win, but bombarding users with irrelevant pop-ups can make them uninstall the app faster than you can say "skip." Studies suggest that intrusive ads are among the top reasons for app churn (Liang et al., 2024; Dogra et al., 2021). Subscriptions, on the other hand, can be a blessing or a curse. If the content doesn't justify the price, users will cancel—and not quietly. Bad reviews can spread like wildfire. In-app purchases? Great, but only if they feel optional rather than essential, avoiding the dreaded "pay-to-win" trap.

Then there's the tech side. Overloading an app with ad trackers or payment gateways can slow it down, leaving users frustrated (He et al., 2024; Fatima-Tuz-Zahra et al., 2020). A sluggish app isn't just annoying; it's a dealbreaker. Meanwhile, the choice of a sustainable model can shape an app's growth. Developers who focus on balanced strategies—ads that respect user time or freemium models with genuine value—tend to fare better in retaining users and building loyalty (Hussain et al., 2024; Sindiramutty, 2024).

In the end, a thoughtful monetization plan isn't just about profits. It's about crafting an experience that keeps users coming back, happily swiping and tapping.

3. COMMON MOBILE APP MONETIZATION MODELS

a) Freemium Model

Explanation of the freemium model (free app with paid premium features) and how it is implemented.

Freemium apps are like free samples at a grocery store—you try them out, and if you like them, you pay for the extras. The basic version is free, but premium features require a purchase. Think of Spotify. It lets you enjoy music for free with ads, but premium users get an ad-free experience and offline playback. This model is appealing because people are more likely to try something free without hesitation (Faria et al., 2024; Gopi et al., 2021). Developers often limit the free version in subtle ways, nudging users toward upgrades. For instance, a photo editing app might offer basic filters for free but lock advanced tools behind a paywall. This balance between offering value and encouraging upgrades is tricky but effective. A friend of mine once got hooked on a freemium fitness app—it started with free workouts, then charged for personalized plans. She resisted, then caved after realizing the free plans were getting repetitive.

In-app purchases and subscriptions are common in freemium models, offering flexibility for users. Developers also gather feedback from free users to refine premium features, keeping the cycle alive (Boudreau et al., 2023; Gouda et al., 2022). Despite its challenges, this strategy brings in revenue while keeping a broad user base. Done well, freemium apps can cater to both casual users and those seeking premium experiences (Enache et al., 2023). Figure 3 show types of monetization model. Figure 3 shows type of monetization model

Figure 3. Type of monetization model (What Is Monetization? | Adjust, 2024)

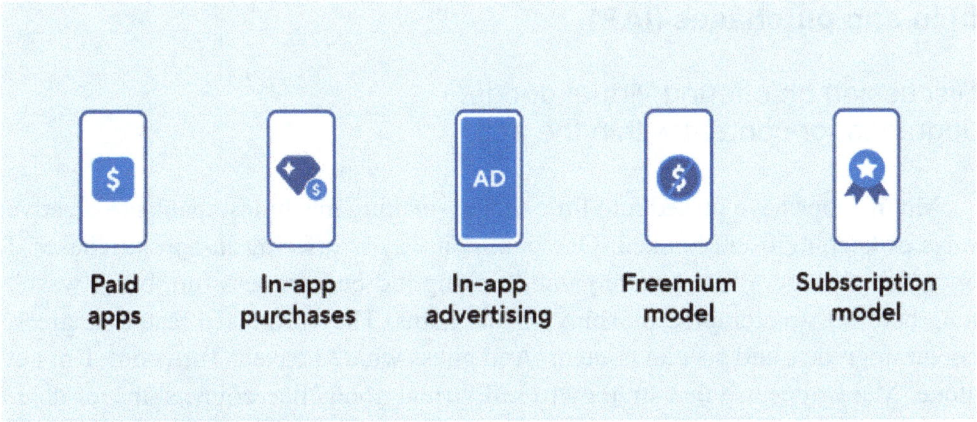

Pros and cons for both users and developers

Freemium apps feel like a win-win at first glance, but they come with trade-offs. For users, the biggest perk is access without an upfront cost. Who wouldn't love that? A friend of mine downloaded a language learning app and used the free version for months, happy as can be. But then she hit a premium-only feature just when she needed it most—cue frustration. That's the catch. The free version is often limited, and ads can get annoying fast (Gui et al., 2015).

For developers, freemium models are a goldmine when done right. Offering a free tier attracts a large audience, giving apps a shot at going viral. More users mean more chances to convert a few into paying customers. But it's a balancing act. If the free features are too good, nobody upgrades. If they're too basic, users leave. And then there's the cost of maintaining free users, which eats into profits if the premium side doesn't deliver.

It's not all bad news. Developers gain valuable user data from the free tier, which helps improve features and target ads (Radivojevic et al., 2024; Humayun et al., 2022). Users benefit, too, from trying an app without the risk of wasting money. The freemium model works best when it respects users' time and offers real value—whether they pay or not (Kowalkowski & Ulaga, 2024).

b) In-app purchases (IAP)

Discussion on offering virtual goods, upgrades, or content within the app.

Mobile apps have turned into little money-making machines, thanks to creative ways of keeping users hooked. One common way is offering in-app purchases. I once downloaded a free puzzle game thinking it'd be harmless fun, but it wasn't long before I was tempted by shiny virtual gems. They promised faster progress, cooler upgrades, and special content. And guess what? I caved. Turns out, I'm not alone. Many apps use this strategy to sell virtual goods like coins, skins, or characters (Putra, 2024).

This isn't just about games either. News apps might charge for premium articles, and fitness apps could hide advanced plans behind a paywall (Malki et al., 2024; Jhanjhi et al., 2021). While it sounds simple, the psychology here is fascinating. People are more likely to spend on something once they've invested time in it. It's why free apps often offer a taste of premium features before asking you to pay. It's like being offered a free coffee sample and suddenly feeling okay about paying $5 for a full cup (Gasiewski, 2024).

The clever part is that these purchases feel personal. You're not buying random extras; you're getting something that makes *your* experience better. However, some users end up overspending, a point often discussed in consumer protection circles. Still, this model has revolutionized how apps turn downloads into dollars.

Analysis of IAP's success in gaming apps and other types of apps.

In-app purchases are everywhere, from addictive games to productivity tools. Gaming apps, though, seem to have mastered the art of persuasion. I once downloaded a popular battle game that was free at first. But then, I noticed my competitors flaunting cool armour and faster weapons—items conveniently available for purchase. Within a week, I spent on virtual coins. This kind of "freemium" model works because it plays on emotions like competitiveness and instant gratification (Shu-Chen, 2023).

But it's not just games cashing in. Lifestyle apps are catching up. Fitness apps, for example, might offer basic features for free but lock personalized plans behind a paywall (Guluzade & Sas, 2024; Kumar et al., 2021). Meditation apps do something similar—unlocking exclusive sessions or sleep tracks for a fee (Li et al., 2024; Lim et al., 2019). What's clever is that these purchases feel like investments in self-improvement. Who doesn't want better health or a good night's sleep?

The success of in-app purchases often depends on how seamlessly they're integrated. Nobody wants to feel like they're being forced to pay. Instead, it's about creating a natural progression where spending feels optional but highly rewarding. That said, the model isn't perfect. Critics argue it can lead to over-reliance or overspending, especially in games designed to tempt impulsive buys. Still, its adaptability across app types is impressive.

c) Subscription Model

Overview of subscription-based monetization (weekly, monthly, or annual)

When it comes to mobile app monetization, subscription-based models are becoming a favorite for many developers (Yan et al., 2023). Whether it's weekly, monthly, or annual, these models create a steady stream of revenue by charging users regularly. As someone who's worked with a few mobile app projects, I can say that this approach offers a balance between profitability and user retention. Apps like Spotify and Netflix have made subscriptions an easy and popular choice, making users feel like they're getting value in exchange for their regular payments (Echauri, 2023).

However, the key challenge is finding the sweet spot for pricing. A weekly subscription can offer low entry costs, which is attractive, but it may annoy users with frequent renewals (Delfianto & Chandra, 2023; Nayyar et al., 2021). On the other hand, monthly and yearly options provide more stability and could even offer users some discounts, encouraging them to stick around longer (Monteiro et al., 2023). I've seen developers get creative with offering limited-time free trials or sneak peeks to draw people in, which is often effective (Dinata et al., 2024).

The subscription model works best when there's a consistent value proposition, keeping users engaged enough to make those recurring payments feel worth it (Hochstein et al., 2023). If the app doesn't deliver on its promises, users are likely to cancel subscriptions quickly, making user feedback crucial in tweaking the experience. The monetization strategy has to strike the right balance to keep users happy, and that's where the real challenge lies.

Evaluation of popular subscription apps (e.g., streaming services, fitness apps).

Subscription-based apps have taken over many aspects of our digital lives, from entertainment to fitness (Rendell et al., 2024; Shah et al., 2022). The model offers users flexibility and consistency, but how well do these popular apps perform? Let's look at a couple of big names like Netflix and Peloton. These services are often cited as examples of how well subscriptions can be integrated into an app's design and user experience (Bergauer, 2024).

Take Netflix, for instance. With millions of global subscribers, it has mastered the art of keeping its audience engaged through a vast library of content that caters to nearly every interest (Okumuş, 2022). The monthly subscription model feels like a fair trade for access to top-tier shows and movies, but it's not just about the content. Their seamless user interface, personalized recommendations, and the ability to pause and resume watching across devices make it almost irresistible to cancel.

On the fitness side, Peloton has become synonymous with home workouts (Rutledge et al., 2024). Their subscription model offers both live and on-demand classes, which can feel like a personalized gym experience at home (Zhang & Sun, 2023; Azam, Dulloo, Majeed, Wan, Xin, & Sindiramutty, 2023). I've used it, and honestly, something is motivating about paying for that class — it adds value and a bit of accountability. But the real selling point isn't just the app itself, but the community feel it creates. While the hardware (bike or treadmill) is expensive, the subscription keeps users coming back with new classes every day.

However, the downside? Some users feel the subscription cost doesn't justify the service if they don't use it consistently, which makes churn a bit higher than expected. Despite this, both Netflix and Peloton have proven that when the value is right, subscribers will stick around.

d) Advertising-Based Model

Types of in-app advertisements (banner ads, interstitials, native ads).

Mobile app developers are constantly looking for ways to make money from their creations, and one of the most common methods is through advertising. The advertising-based model relies on showing ads within the app to generate revenue. It's like the modern-day version of having a billboard in a popular spot – except it's on your phone, in your app. I remember using a weather app once, and I couldn't check the forecast without seeing a banner ad for something completely unrelated, like a new video game.

In-app advertisements come in various types. Banner ads are one of the simplest; they're rectangular ads that appear at the top or bottom of the screen. These are common in free apps and are usually not too intrusive. Interstitial ads, on the other hand, are full-screen ads that pop up between transitions in the app (Lüthy, 2024; Sharma et al., 2021). They can be annoying at times, especially when they block content you're trying to view. Then, there are native ads. These blend with the content of the app, so they don't feel out of place, like when you see a sponsored post on social media.

The challenge for developers is balancing the user experience with the need to make money. Too many ads can drive users away, while too few can leave money on the table. Still, this model is one of the top ways apps make money. If done right, ads don't just bring in cash – they also add a layer of engagement, making the app feel more interactive.

Pros and cons, and how to assess ad performance

The advertising-based model in mobile apps has its advantages and disadvantages. One of the main benefits is that it provides a steady stream of income without the user needing to pay directly for the app (X. Li et al., 2023). This allows the app to be free, which can attract a wider audience. I remember using a game app that I enjoyed, and the ads were a small price to pay for all the free content. However, the downside is that too many ads can be intrusive and annoying, which might drive users to look for alternative apps (Rege & Dimitrova, 2024).

Assessing ad performance is crucial for developers to understand how well their ads are working and where improvements can be made. Metrics like click-through rate (CTR), conversion rate, and eCPM (effective cost per thousand impressions) are commonly used to gauge success (Dong et al., 2023; Singhal et al., 2020). For example, I noticed in one of my apps that banner ads with a higher eCPM were more successful than those with a lower CTR. This shows that just getting clicks isn't enough; the quality of those clicks matters too.

To optimize ad performance, it's essential to test different ad formats and placements to see what works best for your audience. Native ads might blend better and get more engagement, while interstitial ads can be effective if used sparingly (Lacko et al., 2024). Also, keeping track of user feedback can give insights into what they find annoying or useful about the ads, which helps in refining the strategy (Natalie et al., 2024).

4. FACTORS INFLUENCING THE CHOICE OF MONETIZATION STRATEGY

a) Target Audience

How user demographics and behaviours impact the monetization strategy (e.g., age, location, interests).

When choosing a monetization strategy, businesses must consider the target audience carefully. A simple example from my experience is when I tried different subscription models for a project and found that age played a huge role. Younger audiences preferred one-time payments, while older ones leaned towards recurring options. The way people engage with content also matters. For instance, location can influence ad placement, as a user in the U.S. might respond differently to an ad than someone in India, given their different purchasing habits and needs (Pratama, 2024). Beyond location, interests shape what users are willing to pay for. A fitness app I used monetized differently based on whether users were beginners or experts in working out.

Age, location, and interests influence what's likely to grab attention and make people want to spend money. Take the rise of social media influencers: younger users tend to interact more with creators, leading brands to leverage these interactions for sponsored content (Cheng & Zhang, 2024). On the flip side, older audiences might prefer more traditional ad models or product placements (Youn et al., 2023;

Azam, Dulloo, Majeed, Wan, Xin, Tajwar, et al., 2023). Even small differences in behaviours, like how often users check their phones or the times of day they do it, can make a big impact on how businesses choose to make money (Lu et al., 2024).

Understanding how user demographics and behaviours play into monetization is essential. It's not just about making money; it's about finding the right fit for your audience and their specific needs. Figure 4 shows factors influencing the choice of monetization strategy.

Figure 4. Factors Influencing the Choice of Monetization Strategy

b) Market Trends

Current trends in mobile app monetization, include the rise of subscription models and ad-based revenue.

Choosing how to make money from a mobile app feels a bit like picking a dessert at a crowded bakery—you know what you want, but trends and choices keep nudging you toward something else. Mobile app developers face a similar dilemma, navigating through user preferences, competition, and what's currently bringing in cash. Right now, subscription models are standing out. Think Spotify, Netflix, or even those cheeky fitness apps that charge monthly to remind you to exercise—

subscriptions offer reliable income and keep users hooked (Bergauer, 2024b; Azam, Tajwar, Mayhialagan, Davis, Yik, Ali, et al., 2023)

Ad-based revenue, though, isn't fading quietly into the background (Jelinski, 2024). From free games bombarding you with ads every two minutes to cleverly placed native advertising, it's still a favourite, especially for apps targeting wider audiences. But here's the twist: users are growing picky about what they're willing to tolerate. Annoy them with too many ads, and they're gone.

I recently chatted with a friend who develops indie games. She swears by the freemium model—a combination of ads and optional subscriptions. According to her, users are more willing to pay once they've fallen for the app (Al-Shamaileh & Sutcliffe, 2022; Azam, Tan, Pin, Syahmi, Qian, Jingyan, et al., 2023). And she's not alone. Market reports back this up, highlighting the need to balance monetization strategies with user experience (Okeyo, 2024).

In the end, the "right" choice is a mix of creativity and strategy—like finding that dessert that hits the sweet spot.

c) User Experience and Engagement

Balancing monetization with maintaining a positive user experience.

Choosing a monetization strategy often feels like juggling—drop one ball, and you risk losing users; grab too many, and they might leave anyway. I've seen this firsthand in apps I've used where aggressive ads ruined an otherwise enjoyable experience. Striking a balance between making money and keeping users happy isn't easy, but it's crucial. Research shows that user experience directly impacts retention; if people feel bombarded by ads or pressured into spending, they'll leave (McKinnon, 2023). Nobody likes to feel like a wallet with legs.

On the flip side, engagement can flourish when monetization aligns with user needs. Take the success of subscription-based apps that offer ad-free options—users happily pay for a cleaner interface and uninterrupted service. Companies like Spotify and YouTube nailed this approach by catering to their audience's priorities. That said, even these models can backfire if they seem exploitative or overpriced.

Transparency matters too. If users understand how their payments support the service, they're more likely to feel involved rather than exploited. Data also suggests that limited, well-placed ads are far less likely to disrupt user engagement (Fujak et al., 2022; Hussain et al., 2024). Still, the challenge remains—how far can monetization go without crossing the line? The answer, it seems, lies in listening to users and valuing their trust as much as their wallets.

5. EVALUATING THE EFFECTIVENESS OF MONETIZATION STRATEGIES

a) Metrics and KPIs for Monetization Success

Key performance indicators (KPIs) to measure monetization success (e.g., ARPU, LTV, churn rate).

Monetization strategies are all about figuring out what works to turn ideas into revenue. It's not just a numbers game—it's about understanding what those numbers mean. Take **ARPU (Average Revenue Per User)** (Shukla et al., 2024), for example. I remember an app I loved, where ARPU shot up because they introduced a subscription tier I grudgingly paid for (no ads!). But guess what? Not everyone stayed. That's where the **churn rate** came in, showing how fast users left. It's like a revolving door: you want more people coming in than leaving.

Then there's **LTV (Lifetime Value)**, which is the big picture of how much value a user brings over time (Zhou et al., 2024), I think about my Spotify subscription—I've stuck around for years, probably making them more than a casual user ever would. But if they raised prices too much? Bye-bye, loyalty. This is why tracking **conversion rates** is key too. Are free users upgrading? If not, something's off.

Each of these KPIs tells a story, and they're all connected. A drop in LTV might hint at higher churn or users feeling underwhelmed. It's a balancing act, and sometimes, even the best plans flop. But learning from the flops? That's priceless.

b) A/B Testing and Experimentation

Importance of testing different monetization strategies to determine what works best.

Monetizing a product or service effectively can feel like navigating through a maze. There are countless strategies but testing them is essential to figure out what truly clicks with the audience. A/B testing and experimentation are key tools for this. A/B testing involves comparing two versions of a product to see which performs better (Larsen et al., 2023; Jun et al., 2024). It's like offering two flavours of ice cream and seeing which one your friends like more (simple, right?). Experimentation pushes it further by allowing for multiple changes and testing those variations simultaneously. This method isn't just for tech startups—it can help any business refine its strategy and boost revenue.

By testing different monetization methods, businesses can adjust pricing, features, or even ad placements based on real data (Ofulue & Benyoucef, 2022; Manchuri et al., 2024). For instance, a company could test a subscription model versus a pay-per-use option. The results can reveal which is more appealing to users. Moreover, continuous experimentation helps optimize long-term profits, as patterns emerge from various trials (Ros et al., 2023; Ravichandran et al., 2024). There's a certain thrill in seeing how small tweaks impact results. It's like tuning an instrument to perfection, only in business. But the process isn't always smooth failure is just as common as success, and that's part of the learning curve. Being ready to adapt quickly, based on what the experiments reveal, is the goal. Figure 5 shows evaluating the effectiveness of monetization strategies.

Figure 5. Evaluating the Effectiveness of Monetization Strategies

Examples of A/B testing in mobile apps.

A/B testing is a game-changer for mobile app developers. It's like being able to try out different outfits before the big day to see which one gets the most compliments. Apps can experiment with everything from button colours to subscription models. For example, one well-known music app tested two different subscription plans: one with a free trial and another with a lower starting price. By monitoring user engagement and conversion rates, they discovered that the lower-priced plan worked better, driving more sign-ups (Rabhan, 2013; Seng et al., 2024).

In another case, a mobile e-commerce app changed the placement of the "Buy Now" button—moving it higher up on the screen to see if it would lead to more purchases. The experiment showed a noticeable increase in conversions, proving that small design changes could have a big impact on revenue.

Even something as simple as push notifications can be tested. A fitness app, for example, tested two types of push notifications—one with personalized messages and one with general updates (Rabbi et al., 2015). The result? Users who received personalized notifications were more likely to engage with the app and hit their fitness goals (Chen et al., 2024; Sindiramutty, Prabagaran, Akbar, Hussain, & Malik, 2024). For a game app, A/B testing might involve testing different in-game rewards to see what motivates users to keep playing (Châtel et al., 2024). These examples show how A/B testing can be used to optimize nearly every aspect of a mobile app, ensuring it stays relevant and appealing to users.

Case Studies

In-depth analysis of successful mobile apps and the monetization strategies that led to their success (e.g., Candy Crush, Spotify, TikTok).

Mobile app monetization strategies have seen incredible success with apps like Candy Crush, Spotify, and TikTok, each leveraging unique approaches to turn users into revenue (Ryu, 2024). When I first tried Candy Crush, I was sucked into the cycle of "one more level" and those in-game purchases for extra moves. Little did I know, this was the power of in-app purchases (IAP), which have been a key driver of Candy Crush's success. Users keep playing, spend small amounts frequently, and the revenue flows in. Figure 6 shows successful mobile apps and the monetization strategies.

Figure 6. Successful mobile apps and the monetization strategies

On the other hand, Spotify has capitalized on the freemium model (J. Li, Wu, et al., 2024). I'm one of many who started with a free account, and then upgraded to a premium once the ads drove me crazy. This model allows free users to experience the app and eventually convert to paying customers. The balance between free access and premium perks is something other apps, like TikTok, have also adopted (X. Zhang et al., 2024; Sindiramutty, Prabagaran, Jhanjhi, Ghazanfar, Malik, & Soomro, 2024). TikTok's reliance on ad revenue is another powerful strategy. I still remember the first time I saw an ad pop up between my endless scrolling; it didn't feel intrusive, which shows how well-targeted these ads are.

What's interesting is the psychological aspect of these strategies. Users often feel like they've invested time, and small purchases don't seem like a big deal. As apps like TikTok get more personalized, they push ads based on behaviour and preferences, making monetization even smoother (Nowacki, 2024). These strategies don't just keep users engaged—they also turn engagement into cash.

6. CHALLENGES AND PITFALLS IN MOBILE APP MONETIZATION

a) User Resistance to Ads and In-App Purchases

Discussion on how users perceive ads and microtransactions, and how to mitigate negative feedback.

Mobile app monetization, particularly through ads and in-app purchases, faces significant challenges due to user resistance (Paschmann et al., 2024). Many users express frustration with intrusive ads, feeling that they disrupt their experience. I've personally encountered this when trying out a free game that bombarded me with ads every few minutes. It felt less like a fun break and more like a sales pitch. This reaction isn't unique; studies show that users tend to abandon apps with excessive ads, especially if they aren't integrated smoothly into the app's flow (Morais, 2024).

Similarly, in-app purchases, like microtransactions for extra lives or premium features, often frustrate users who feel pressured to pay for features that should be accessible (Sellgren, 2024; Sindiramutty, Prabagaran, Jhanjhi, Murugesan, Brohi, & Masud, 2024). These purchases can lead to negative feedback, with users labelling them as "pay-to-win" schemes, damaging the app's reputation.

To mitigate this, it's essential to find a balance. Offering ad-free versions for a reasonable price or letting users skip ads after watching for a set time can soften the blow (K. Zhang et al., 2024). Similarly, instead of forcing in-app purchases, providing value through optional upgrades can help retain users without turning them away. This was the case with "Candy Crush," which allowed users to enjoy the game for free but gave them the option to enhance their experience with purchases (Moradzadeh & Kou, 2024). Figure 7 shows challenges and pitfalls in mobile app monetization.

Figure 7. Challenges and Pitfalls in Mobile App Monetization

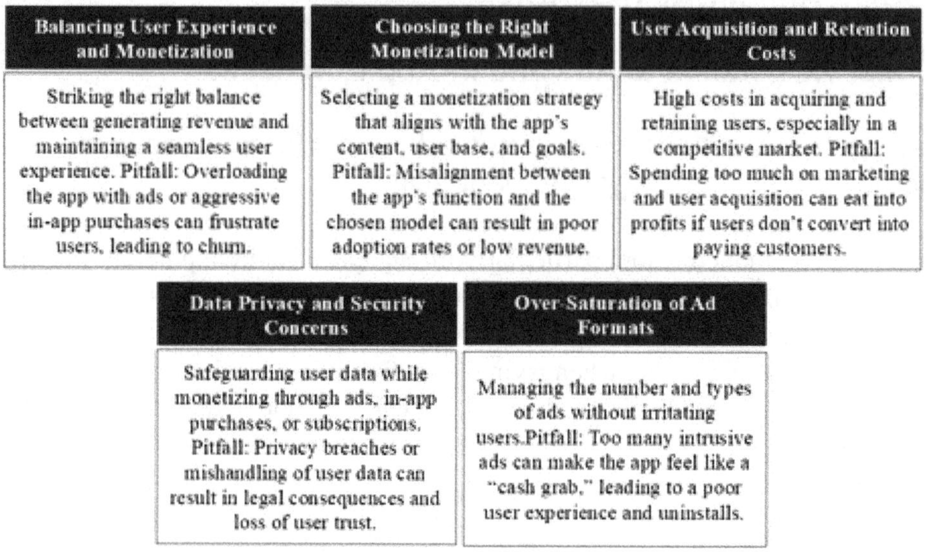

Addressing user feedback and making monetization more transparent is key to maintaining a loyal user base (Nwaimo et al., 2024). Listening to complaints and adapting strategies can turn a negative experience into one that users are willing to support.

b) App Store Guidelines and Regulations

Challenges posed by app store policies (e.g., Apple's App Store and Google Play Store).

Monetizing mobile apps can feel like a constant balancing act, especially with app store regulations that often change without warning (Kircher & Foerderer, 2023; Sindiramutty, Prabagaran, Jhanjhi, Murugesan, Brohi, & Wei, 2024). Take Apple's App Store and Google Play, for example. Both platforms have strict rules that developers must follow, which can be frustrating. For instance, Apple takes a hefty commission on in-app purchases (up to 30%), which makes it hard for developers to turn a profit (Tirole, 2024). Google Play isn't much different, with its own set of rules that sometimes seem more like obstacles than guidelines.

I've had friends who launched their apps only to realize that these policies meant unexpected fees or limitations. One friend, who created a fitness tracking app, found that Apple's guidelines restricted the use of third-party payment processors. This meant his app couldn't use cheaper alternatives, forcing him to raise subscription prices. As a result, users weren't too thrilled, and his app's success was cut short.

Then there's the issue of app review processes, which can be a black box. Developers often don't get clear reasons for rejections, leaving them scrambling to figure out how to fix issues (Z. S. Li, Arony, et al., 2024; Sindiramutty, Prabagaran, Jhanjhi, Murugesan, Malik, & Hussain, 2024). This lack of transparency can slow down innovation and stifle growth. Plus, strict data privacy policies can add another layer of complexity, especially for apps that rely on user data for personalized services (Chukwunweike et al., 2024).

These challenges are real, and sometimes it feels like app store policies are more of a roadblock than a help. Still, some developers manage to succeed by navigating these tricky waters and adapting quickly.

c) Maintaining Long-Term Profitability

Addressing the risks of relying too heavily on one monetization method (e.g., ad fatigue or subscription cancellations).

Mobile app developers face a serious challenge when trying to maintain long-term profitability (Rodrik & Sandhu, 2024; Weiqi et al., 2024). Relying on one monetization method, like ads or subscriptions, can lead to issues like ad fatigue or subscription cancellations (Lange, 2024). For instance, apps like *Candy Crush* once thrived on in-app purchases, but many users eventually stopped buying boosts, diminishing revenue. Similarly, subscription models can seem promising, but cancellations are a common risk (Sheil et al., 2024; Ying et al., 2024). Apps such as *Spotify* and *Netflix* have seen fluctuations in subscribers as users juggle different services.

What happens is that users can grow tired of frequent ads, or they may not want to keep paying for a service they no longer use. This is where balance is key—relying too much on ads can ruin the user experience, while too many subscription fees might scare users away. For example, *TikTok* generates massive revenue from ads but still struggles to keep users engaged without overloading them with advertisements (Dahl, 2024).

To avoid these pitfalls, successful apps often diversify their monetization methods. *Candy Crush* added advertising as a secondary income stream, while *Spotify* has a free version to lure in users who might later convert to paying subscribers (K et al., 2024). Apps need to be flexible, constantly adjusting their strategies to ensure they don't alienate users while still making a profit.

7. FUTURE TRENDS IN MOBILE APP MONETIZATION

a) Emerging Monetization Models

Exploration of new and experimental monetization models (e.g., blockchain-based models, NFTs in mobile apps).

The future of mobile app monetization is shifting toward more creative and experimental models. Traditional ad-based strategies are still prevalent, but innovative approaches are gaining traction. Blockchain technology is emerging as a game-changer in this space. Imagine, for example, apps allowing users to earn cryptocurrency or trade in-game assets on a blockchain, creating a new type of virtual economy. Non-fungible tokens (NFTs) are also finding their way into mobile apps, offering exclusive digital assets or rewards for loyal users (Varadarajan & Seeni, 2024). This can be seen in games like *Gods Unchained*, where players buy and sell NFTs. Figure 8 shows future trends in mobile app monetization.

Figure 8. Future Trends in Mobile App Monetization

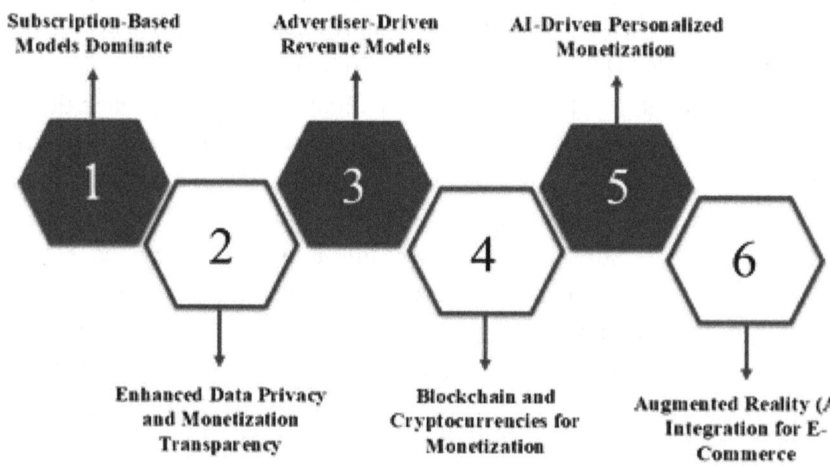

These new monetization strategies are more than just gimmicks; they offer real value to users, making the experience more engaging. Take *TikTok*, for instance—its "gifts" feature allows users to purchase virtual items for creators, creating a thriving micro-economy within the app. And then there's *Spotify*, which has experimented with subscription models offering not just music but exclusive experiences for premium members, a win-win for both users and creators (Suhod et al., 2024).

What's exciting is how these models aren't just about making money—they are reshaping how apps interact with users, creating a sense of ownership and investment. Blockchain and NFTs have opened the door to more personalized, user-driven experiences that could be the next big thing in mobile app monetization (Pizzo et al., 2024; Waheed et al., 2024).

b) Personalization and AI in Monetization

How artificial intelligence and data analytics are shaping the future of personalized monetization (e.g., targeted ads, predictive subscriptions).

Mobile app monetization is shifting rapidly, thanks to advances in artificial intelligence (AI) and data analytics. Apps today are personalizing how they make money, and it's all about knowing the user. I've seen this firsthand with my favorite apps like Spotify, where personalized recommendations are not just about music, but about subscription plans too. AI is making it possible for apps to offer targeted ads that feel less intrusive and more relevant. The app learns what you like and shows you ads based on your preferences, making ads feel more like a suggestion than an interruption (Hsu & Karahalios, 2024; Wen et al., 2023).

Predictive analytics is also helping businesses understand when a user might be ready to subscribe, or when to offer a discount to keep them engaged (Ijomah et al., 2024).

Take TikTok, for instance, which thrives by offering personalized experiences, not just through content but monetization too. Their algorithm doesn't just show you viral videos; it also knows when you're most likely to engage with in-app purchases or ads, based on your behaviour (Ringborg & Ohlsson, 2024). And don't forget how app developers use AI to segment users more precisely than ever, which leads to higher conversion rates for both in-app purchases and subscriptions (Patil et al., 2024). AI and data are shaping the future of monetization, offering businesses smarter ways to make money while keeping users engaged. It's a win-win, and the apps that get it right, like Candy Crush, show us how it's done.

8. CONCLUSION

• Summary of Key Points

Recap of the main monetization models and the criteria for choosing the right strategy.

When it comes to turning mobile apps into revenue-generating tools, the strategies developers choose can make or break their success. Monetization models include freemium, in-app purchases, ads, and subscriptions. Each has its quirks and works differently depending on the app type and audience.

Freemium models hook users with free features but charge for extras—think Spotify's ad-free tier. It's a clever balance: too many freebies, and users won't upgrade; too few, and they leave. In-app purchases are the bread and butter for games like Candy Crush, offering players extras to enhance their experience. Ads remain popular for free apps, but a flood of poorly placed pop-ups can scare users away faster than you can hit "skip." Subscriptions, like Netflix's recurring payments, ensure steady income but need constant updates to keep users interested.

Choosing the right approach is all about knowing your audience. Are they casual users who'll tolerate ads, or die-hard fans ready to pay for premium features? Real-world cases highlight this: Spotify converted annoyed free users into subscribers, while Candy Crush monetized through microtransactions without overwhelming players. Meanwhile, apps that misjudge user expectations—like pushing a subscription when one-time purchases suffice—risk backlash and lost loyalty.

The secret sauce? Experiment, adapt, and listen. Successful apps don't just slap on a monetization model; they fine-tune strategies to match user needs while staying profitable. It's a high-stakes game, but for those who get it right, the rewards are worth the hustle.

Final Thoughts

Reflection on the dynamic nature of mobile app monetization and the need for continuous assessment and adaptation.

The world of mobile app monetization is anything but static. What works today might be obsolete tomorrow, and that's both the challenge and the thrill. I remember a developer friend who was riding high on ad revenue from his casual gaming app until user complaints about intrusive ads started rolling in. He pivoted to a freemium model, added optional in-app purchases, and saw a surge in user retention. The lesson? Adaptation isn't optional—it's survival.

Mobile app monetization is like a living organism. Trends shift, user preferences evolve, and competition gets fiercer by the day. Just when developers think they've nailed their strategy, a disruptor like AI-powered personalization or blockchain-based rewards changes the game. For instance, AI has made monetization smarter by tailoring ads and subscriptions to individual users. But with every leap forward, new hurdles emerge, like balancing privacy with personalization or addressing user fatigue.

Success in this space demands continuous assessment. Developers must test, tweak, and even scrap strategies to stay relevant. Metrics like churn rates and lifetime value (LTV) aren't just numbers—they're clues about what's clicking and what's not. It's about staying curious, listening to feedback, and being brave enough to pivot when needed.

In the end, the only constant in mobile app monetization is change. The most successful apps are those that embrace this fluidity, treating each challenge as an opportunity to innovate and connect with users in meaningful ways.

REFERENCES

Akhil. (2024, July 4). *How to Monetize your Mobile App: a guide for entrepreneurs*. Apptunix Blog. https://www.apptunix.com/blog/mobile-app-monetization/

Al-Shamaileh, O., & Sutcliffe, A. (2022). Why people choose Apps: An evaluation of the ecology and user experience of mobile applications. *International Journal of Human-Computer Studies*, *170*, 102965. DOI: 10.1016/j.ijhcs.2022.102965

Alex, S. A., Jhanjhi, N., Humayun, M., Ibrahim, A. O., & Abulfaraj, A. W. (2022). Deep LSTM Model for Diabetes Prediction with Class Balancing by SMOTE. *Electronics (Basel)*, *11*(17), 2737. DOI: 10.3390/electronics11172737

Alferidah, D. K., & Jhanjhi, N. (2020). Cybersecurity Impact over Bigdata and IoT Growth. *2020 International Conference on Computational Intelligence (ICCI)*. DOI: 10.1109/ICCI51257.2020.9247722

Alkinani, M. H., Almazroi, A. A., Jhanjhi, N., & Khan, N. A. (2021). 5G and IoT Based Reporting and Accident Detection (RAD) System to Deliver First Aid Box Using Unmanned Aerial Vehicle. *Sensors (Basel)*, *21*(20), 6905. DOI: 10.3390/s21206905 PMID: 34696118

Ananna, F. F., Nowreen, R., Jahwari, S. S. R. A., Costa, E. A., Angeline, L., & Sindiramutty, S. R. (2023). Analysing Influential factors in student academic achievement: Prediction modelling and insight. *International Journal of Emerging Multidisciplinaries Computer Science & Artificial Intelligence*, *2*(1). Advance online publication. DOI: 10.54938/ijemdcsai.2023.02.1.254

Ashok, P., & Hallur, G. (2024). Navigating the Infotainment Landscape: Status and Regulations in the Digital Age. *2024 International Conference on Advances in Data Engineering and Intelligent Computing Systems (ADICS)*, 1–6. DOI: 10.1109/ADICS58448.2024.10533560

Azam, H., Dulloo, M. I., Majeed, M. H., Wan, J. P. H., Xin, L. T., & Sindiramutty, S. R. (2023). Cybercrime Unmasked: Investigating cases and digital evidence. *International Journal of Emerging Multidisciplinaries Computer Science & Artificial Intelligence*, *2*(1). Advance online publication. DOI: 10.54938/ijemdcsai.2023.02.1.255

Azam, H., Dulloo, M. I., Majeed, M. H., Wan, J. P. H., Xin, L. T., Tajwar, M. A., & Sindiramutty, S. R. (2023). Defending the digital Frontier: IDPS and the battle against Cyber threat. *International Journal of Emerging Multidisciplinaries Computer Science & Artificial Intelligence*, *2*(1). Advance online publication. DOI: 10.54938/ijemdcsai.2023.02.1.253

Azam, H., Tajwar, M. A., Mayhialagan, S., Davis, A. J., Yik, C. J., Ali, D., & Sindiramutty, S. R. (2023). Innovations in Security: A study of cloud Computing and IoT. *International Journal of Emerging Multidisciplinaries Computer Science & Artificial Intelligence, 2*(1). Advance online publication. DOI: 10.54938/ijem-dcsai.2023.02.1.252

Azam, H., Tan, M., Pin, L. T., Syahmi, M. A., Qian, A. L. W., Jingyan, H., Uddin, M. F., & Sindiramutty, S. R. (2023). Wireless Technology Security and Privacy: A Comprehensive Study. *Preprints.org*. DOI: 10.20944/preprints202311.0664.v1

Babbar, H., Rani, S., Masud, M., Verma, S., Anand, D., & Jhanjhi, N. (2021). Load balancing algorithm for migrating switches in software-defined vehicular networks. *Computers, Materials & Continua/Computers. Computers, Materials & Continua, 67*(1), 1301–1316. DOI: 10.32604/cmc.2021.014627

Baek, S., Ahn, J., & Kim, D. (2023). Future Business Model for Mobile Cloud Gaming: The Case of South Korea and Implications. *IEEE Communications Magazine, 61*(7), 68–73. DOI: 10.1109/MCOM.001.2200374

Bapna, R., & Ghose, A. (2024). *Thrive: Maximizing Well-Being in the Age of AI.* MIT Press. DOI: 10.7551/mitpress/15438.001.0001

Bergauer, A. (2024a). *Run it like a business: Strategies for Arts Organizations to Increase Audiences, Remain Relevant, and Multiply Money—Without Losing the Art.* BenBella Books.

Bergauer, A. (2024b). *Run it like a business: Strategies for Arts Organizations to Increase Audiences, Remain Relevant, and Multiply Money—Without Losing the Art.* BenBella Books.

Boudreau, K. J., Jeppesen, L. B., & Miric, M. (2023). Free(mium) strategies for digital goods. In *Edward Elgar Publishing eBooks* (pp. 126–141). DOI: 10.4337/9781800378902.00012

Brohi, S. N., Jhanjhi, N., Brohi, N. N., & Brohi, M. N. (2020). Key Applications of State-of-the-Art Technologies to Mitigate and Eliminate COVID-19.pdf. *TECHRxiv*. DOI: 10.36227/techrxiv.12115596.v1

Châtel, B. D. L., Janssen, J. H. M., Peeters, G. M. E. E., Corten, R., Tieben, R., Deen, M., Hendriks, E. J. M., & Rikkert, M. G. M. O. (2024). Social Gaming to Decrease loneliness in Older Adults: Recruitment Challenges and attrition Analysis in a digital Mixed Methods Feasibility study. *JMIR Serious Games, 12*, e52640. DOI: 10.2196/52640 PMID: 39412863

Chen, Y., Woodward, J., Bista, D., Zhang, X., Singh, I., Obajemu, O., Shankar, M. N., Ross, K. M., Ruiz, J., & Anthony, L. (2024). *Investigating Contextual Notifications to Drive Self-Monitoring in mHealth Apps for Weight Maintenance*. ACM., DOI: 10.1145/3613904.3641993

Cheng, M., & Zhang, S. (2024). Reputation Burning: Analyzing the impact of brand sponsorship on social influencers. *Management Science*, mnsc.2023.00193. Advance online publication. DOI: 10.1287/mnsc.2023.00193

Chesti, I. A., Humayun, M., Sama, N. U., & Jhanjhi, N. (2020). Evolution, Mitigation, and Prevention of Ransomware. *2020 2nd International Conference on Computer and Information Sciences (ICCIS)*. DOI: 10.1109/ICCIS49240.2020.9257708

Chukwunweike, N. J. N., Yussuf, N. M., Okusi, N. O., Bakare, N. T. O., & Abisola, N. J. (2024). The role of deep learning in ensuring privacy integrity and security: Applications in AI-driven cybersecurity solutions. *World Journal of Advanced Research and Reviews*, *23*(2), 1778–1790. DOI: 10.30574/wjarr.2024.23.2.2550

Dahl, H. (2024). *How to Create Value in Social Media: a Postphenomenological Perspective : The technological mediation of social media between users and business - the effect on value of human factors engineering and customer engagement seen from a postphenomenological view*. DIVA. https://www.diva-portal.org/smash/record.jsf?pid=diva2%3A1882279&dswid=8584

Delfianto, H. F., & Chandra, Y. U. (2023). The Impact of Key Factors on Subscription Service Adoption and Retention in Indonesia. *2023 International Conference on Computer Science, Information Technology and Engineering (ICCoSITE)*, 909–914. DOI: 10.1109/ICCoSITE57641.2023.10127749

Dinata, H., Suryani, E., & Purnomo, J. D. T. (2024). Exploring the key success strategy for mobile game developers: A distinguish between premium and freemium versions. *Edelweiss Applied Science and Technology*, *8*(6), 1096–1108. DOI: 10.55214/25768484.v8i6.2213

Djaruma, H., Wirawan, Z. W., Gunawan, A. S., & Saputra, K. E. (2023). Systematic Literature Review of Monetization in Mobile App: Strategies, Trends, Challenges, and Best Practices. *2023 IEEE International Conference on Communication, Networks and Satellite (COMNETSAT)*. DOI: 10.1109/COMNETSAT59769.2023.10420748

Dogra, V., Singh, A., Verma, S., Kavita, N., Jhanjhi, N. Z., & Talib, M. N. (2021). Analyzing DistilBERT for Sentiment Classification of Banking Financial News. In *Lecture notes in networks and systems* (pp. 501–510). DOI: 10.1007/978-981-16-3153-5_53

Dong, B., Zhuang, M., Fang, E., & Huang, M. (2023). Tales of two channels: Digital advertising performance between AI recommendation and user subscription channels. *Journal of Marketing*. Advance online publication. DOI: 10.1177/00222429231190021

Drott, E. (2024). Surveillance of and through Music. In *Surveillance of and through Music* (pp. 273–289). DOI: 10.1007/978-3-031-64013-1_17

Dudzinskaite, U., Correia, R., Venciute, D., & Fontes, R. (2024). Increasing customer engagement in digital marketing campaigns in a time of AI. In *Advances in marketing, customer relationship management, and e-services book series* (pp. 48–80). DOI: 10.4018/979-8-3693-2153-9.ch003

Echauri, G. (2023). Digital subscribers: Between freedom and constraint. *Media Culture & Society, 45*(6), 1175–1191. DOI: 10.1177/01634437231159554

Enache, A., Friberg, R., & Wiklander, M. (2023). Demand for in-app purchases in mobile apps—A difference-in-difference approach. *International Journal of Industrial Organization, 88*, 102945. DOI: 10.1016/j.ijindorg.2023.102945

Faria, A. M., Dias, Á., Pereira, L. F., d, R. L., Costa, , & Gonçalves, R. (2024). Investigating factors that affect willingness to pay: An analysis on freemium social media apps. *International Journal of Advanced Media and Communication, 8*(1), 12–52. DOI: 10.1504/IJAMC.2024.140638

Fatima-Tuz-Zahra, N., Jhanjhi, N., Brohi, S. N., Malik, N. A., & Humayun, M. (2020). Proposing a Hybrid RPL Protocol for Rank and Wormhole Attack Mitigation using Machine Learning. *2020 2nd International Conference on Computer and Information Sciences (ICCIS)*. DOI: 10.1109/ICCIS49240.2020.9257607

Fujak, H., Karg, A., & Whitburn, D. (2022). Sport media has an audience measurement problem. *European Sport Management Quarterly, 24*(3), 582–602. DOI: 10.1080/16184742.2022.2153376

Gasiewski, M. (2024). Revolutionizing revenue at Google. In *Routledge eBooks* (pp. 43–49). DOI: 10.4324/9781003535966-5

Gopi, R., Sathiyamoorthi, V., Selvakumar, S., Manikandan, R., Chatterjee, P., Jhanjhi, N. Z., & Luhach, A. K. (2021). Enhanced method of ANN based model for detection of DDoS attacks on multimedia internet of things. *Multimedia Tools and Applications, 81*(19), 26739–26757. DOI: 10.1007/s11042-021-10640-6

Gouda, W., Almurafeh, M., Humayun, M., & Jhanjhi, N. Z. (2022). Detection of COVID-19 based on chest x-rays using deep learning. *Health Care, 10*(2), 343. DOI: 10.3390/healthcare10020343 PMID: 35206957

Gui, J., Mcilroy, S., Nagappan, M., & Halfond, W. G. J. (2015). Truth in advertising: the hidden cost of mobile ads for software developers. *International Conference on Software Engineering, 1*, 100–110. DOI: 10.1109/ICSE.2015.32

Guluzade, L., & Sas, C. (2024). Functionality and user review analysis of mobile apps for mindfulness eating and eating disorders. *Designing Interactive Systems Conference, 6*, 1350–1371. DOI: 10.1145/3643834.3661521

He, Z., Huq, S. F., & Malek, S. (2024). *"I tend to view ads almost like a pestilence": On the Accessibility Implications of Mobile Ads for Blind Users*. ACM., DOI: 10.1145/3597503.3639228

Hochstein, B., Voorhees, C. M., Pratt, A. B., Rangarajan, D., Nagel, D. M., & Mehrotra, V. (2023). Customer success management, customer health, and retention in B2B industries. *International Journal of Research in Marketing*, *40*(4), 912–932. DOI: 10.1016/j.ijresmar.2023.09.002

Hsu, S., & Karahalios, K. (2024). *Choosing What You Want Versus Getting What You Want: An Experiment with Choice in Video Ad Placement*. ACM., DOI: 10.1145/3613904.3642869

Humayun, M., Sujatha, R., Almuayqil, S. N., & Jhanjhi, N. Z. (2022). A Transfer Learning Approach with a Convolutional Neural Network for the Classification of Lung Carcinoma. *Health Care*, *10*(6), 1058. DOI: 10.3390/healthcare10061058 PMID: 35742109

Hussain, A., Ting, D. H., & Marder, B. (2024). Why premium in freemium: A hedonic shopping motivation model in virtual game retailing. *Information Technology & People*. Advance online publication. DOI: 10.1108/ITP-01-2023-0082

Hussain, K., Rahmatyar, A. R., Riskhan, B., Sheikh, M. U., & Sindiramutty, S. R. (2024). Threats and Vulnerabilities of Wireless Networks in the Internet of Things (IoT). *2024 IEEE 1st Karachi Section Humanitarian Technology Conference (KHI-HTC), 2*, 1–8. DOI: 10.1109/KHI-HTC60760.2024.10482197

Ijomah, N. T. I., Idemudia, N. C., Eyo-Udo, N. N. L., & Anjorin, N. K. F. (2024). The role of big data analytics in customer relationship management: Strategies for improving customer engagement and retention. *World Journal of Advanced Science and Technology, 6*(1), 013–024. DOI: 10.53346/wjast.2024.6.1.0038

Jelinski, J. (2024). *Needle work: A History of Commercial Tattooing in Canada Volume 44*.

Jhanjhi, N., Humayun, M., & Almuayqil, S. N. (2021). Cyber security and privacy issues in industrial internet of things. *Computer Systems Science and Engineering*, *37*(3), 361–380. DOI: 10.32604/csse.2021.015206

Jun, A. Y. M., Jinu, B. A., Seng, L. K., Maharaiq, M. H. F. B. Z., Khongsuwan, W., Junn, B. T. K., Hao, A. W., & Sindiramutty, S. R. (2024). Exploring the Impact of Crypto-Ransomware on Critical Industries: Case Studies and Solutions. *Preprints.org*. DOI: 10.20944/preprints202409.1325.v1

Katariya, J. (2024, June 19). 15 Mobile App Monetization Strategies for Maximizing Revenue. *Moon Technolabs*. https://www.moontechnolabs.com/blog/mobile-app-monetization/

Kircher, T., & Foerderer, J. (2023). Ban Targeted Advertising? An Empirical investigation of the consequences for app development. *Management Science*, *70*(2), 1070–1092. DOI: 10.1287/mnsc.2023.4726

Kiyani, F. F., Hamid, B., Humayun, M., Sindiramutty, S. R. L., & Chowdhury, S. (2024). Discovery of Influential Publications Using Research Article's Usage Context. *2024 International Conference on Emerging Trends in Networks and Computer Communications (ETNCC)*, 1–7. DOI: 10.1109/ETNCC63262.2024.10767576

Kowalkowski, C., & Ulaga, W. (2024). Subscription offers in business-to-business markets: Conceptualization, taxonomy, and framework for growth. *Industrial Marketing Management*, *117*, 440–456. DOI: 10.1016/j.indmarman.2024.01.014

Kumar, M. S., Vimal, S., Jhanjhi, N., Dhanabalan, S. S., & Alhumyani, H. A. (2021). Blockchain based peer to peer communication in autonomous drone operation. *Energy Reports*, *7*, 7925–7939. DOI: 10.1016/j.egyr.2021.08.073

Lacko, D., Machackova, H., & Slavík, L. (2024). Adolescents' perceptions of the credibility of informational content on fitness and dietary supplements: The impact of banner and native advertising. *Journal of Adolescence*, *96*(8), 1956–1968. Advance online publication. DOI: 10.1002/jad.12394 PMID: 39164994

Lange, P. G. (2024). Ranting in emotional public spheres: Publicizing participatory challenges on YouTube. *Convergence (London)*, *30*(2), 904–921. DOI: 10.1177/13548565241246044

Larsen, N., Stallrich, J., Sengupta, S., Deng, A., Kohavi, R., & Stevens, N. T. (2023). Statistical challenges in online controlled experiments: A review of A/B testing methodology. *The American Statistician*, •••, 1–15. DOI: 10.1080/00031305.2023.2257237

Lemos, N., Machado, C. S., & Cardoso, C. (2023). Free apps and paid apps: Monetization strategies for health apps in the Portuguese market. *International Journal of Pharmaceutical and Healthcare Marketing, 18*(2), 237–251. DOI: 10.1108/IJPHM-01-2023-0001

Li, J., Cochrane, K. A., & Leshed, G. (2024). Beyond meditation: understanding everyday mindfulness practices and technology use among experienced practitioners. *Proceedings of the ACM on Human-Computer Interaction, 8*(CSCW2), 1–29. DOI: 10.1145/3687023

Li, J., Wu, J., Zhang, J., & Ji, X. (2024). Freemium design: Optimal tier differentiation models for content platforms. *Transportation Research Part E, Logistics and Transportation Review, 188*, 103612. DOI: 10.1016/j.tre.2024.103612

Li, X., Balasubramanian, H., Chen, Y., & Pang, C. (2023). Managing conflicting revenue streams from advertisers and subscribers for online platforms. *European Journal of Operational Research, 314*(1), 241–254. DOI: 10.1016/j.ejor.2023.10.024

Li, Z. S., Arony, N. N., Awon, A. M., Damian, D., & Xu, B. (2024). AI tool use and adoption in software development by Individuals and Organizations: A Grounded Theory Study. *arXiv (Cornell University)*. https://doi.org//arxiv.2406.17325DOI: 10.48550

Liang, S., Wei, Z., & Zang, L. (2024). An exploratory study of factors influencing user app abandonment on smartphones. *Library Hi Tech*. Advance online publication. DOI: 10.1108/LHT-07-2022-0349

Lim, M., Abdullah, A., Jhanjhi, N., Khan, M. K., & Supramaniam, M. (2019). Link Prediction in Time-Evolving Criminal Network with deep Reinforcement learning technique. *IEEE Access : Practical Innovations, Open Solutions, 7*, 184797–184807. DOI: 10.1109/ACCESS.2019.2958873

Lindgren, E. (2024). *Accessibility redesign and its effect on purchase intent and customer experience*. DIVA. https://www.diva-portal.org/smash/record.jsf?pid=diva2%3A1864084&dswid=-285

Lindström, C. W. J., Vishkaei, B. M., & De Giovanni, P. (2023). Subscription-based business models in the context of tech firms: theory and applications. *International Journal of Industrial Engineering and Operations Management*. DOI: 10.1108/IJIEOM-06-2023-0054

Linqiang, Y., Sindiramutty, S. R. L., Ashraf, H., Muzammal, S. M., Balakrishnan, S. P., Gupta, S., & Kavita, N. (2024). Intelligent Household Waste Classification System Based on Machine Learning. *2024 International Conference on Emerging Trends in Networks and Computer Communications (ETNCC)*, 760–768. DOI: 10.1109/ETNCC63262.2024.10767563

Lu, J., Bradlow, E. T., & Hutchinson, J. W. (2024). More Likely to Pay but Less Engaged: The Effects of Switching Online Courses from Scheduled to On-Demand Release on User Behavior. *Journal of Marketing*, *88*(5), 63–88. DOI: 10.1177/00222429241227145

Lut, L. T. Y. (2024). *Utilizing customer preferences to enhance digital services : appealing features of subscription model and subscription management portal.* LUTPub. https://lutpub.lut.fi/handle/10024/168439

Lüthy, K. (2024). Periodicals/Subject groups. In *Springer eBooks* (pp. 765–789). DOI: 10.1007/978-3-658-39909-2_33

Maglogiannis, I. (2024). *Artificial Intelligence Applications and Innovations. AIAI 2024 IFIP WG 12.5 International Workshops: MHDW 2024, 5G-PINE 2024, and AI4GD 2024, Corfu, Greece, June 27–30, 2024, Proceedings*. Springer Nature.

Malki, L. M., Kaleva, I., Patel, D., Warner, M., & Abu-Salma, R. (2024). *Exploring Privacy Practices of Female mHealth Apps in a Post-Roe World.* ACM., DOI: 10.1145/3613904.3642521

Manchuri, A., Kakera, A., Saleh, A., & Raja, S. (2024). pplication of Supervised Machine Learning Models in Biodiesel Production Research - A Short Review. *Borneo Journal of Sciences and Technology*. DOI: 10.35370/bjost.2024.6.1-10

McKinnon, M. M. (2023). *Ad or Editorial? Native Advertising and Shifting Political Attitudes - ProQuest*. https://www.proquest.com/openview/1a3e962f1720be3b291866b43ff9be2d/1?cbl=18750&diss=y&pq-origsite=gscholar

Mihale-Wilson, C., & Carl, K. V. (2024). Designing incentive systems for participation in digital ecosystems—An integrated framework. *Electronic Markets*, *34*(1), 16. Advance online publication. DOI: 10.1007/s12525-024-00703-5

Monteiro, M. M., De Azevedo, C. M. L., Kamargianni, M., Cantelmo, G., Tavory, S. S., Gal-Tzur, A., Antoniou, C., & Shiftan, Y. (2023). Car-sharing subscription preferences and the role of incentives: The case of Copenhagen, Munich, and Tel Aviv-Yafo. *Case Studies on Transport Policy*, *12*, 101013. DOI: 10.1016/j.cstp.2023.101013

Moradzadeh, S., & Kou, Y. (2024). "Wow another fake game from YouTube ad": Unpacking Fake Games Through a Mixed-Methods Investigation. *Proceedings of the ACM on Human-Computer Interaction, 8*(CHI PLAY), 1–36. DOI: 10.1145/3677115

Morais, R. A. (2024). *Design & Advertising: Applications of design in advertising departments*. Editora Appris.

Natalie, R., Chang, R., Sheshadri, S., Guo, A., & Hara, K. (2024)... *Audio Description Customization. ACM, 84*, 1–19. DOI: 10.1145/3663548.3675617

Nayyar, A., Gadhavi, L., & Zaman, N. (2021). Machine learning in healthcare: review, opportunities and challenges. In *Elsevier eBooks* (pp. 23–45). DOI: 10.1016/B978-0-12-821229-5.00011-2

Norizan, A. R., Kamil, M. H. F. M., Zamri, M. F. M., & Khalid, F. S. (2024). Comparing In-Game purchasing behavior between youth and adult gamers: Understanding enjoyment rates and implications for game developers. In *Studies in systems, decision and control* (pp. 303–316). DOI: 10.1007/978-3-031-62656-2_27

Nowacki, L. (2024). The old king is dead, long live the algorithmic king – the decline of Facebook and the rise of TikTok – comparative study of algorithmic design of social media platforms. *International Social Science Journal, 74*(254), 1325–1338. Advance online publication. DOI: 10.1111/issj.12513

Nwaimo, N. C. S., Adegbola, N. E., & Adegbola, N. M. D.Chioma Susan NwaimoAyodeji Enoch AdegbolaMayokun Daniel Adegbola. (2024). Data-driven strategies for enhancing user engagement in digital platforms. *International Journal of Management & Entrepreneurship Research, 6*(6), 1854–1868. DOI: 10.51594/ijmer.v6i6.1170

Ofulue, J., & Benyoucef, M. (2022). *Data monetization: insights from a technology-enabled literature review and research agenda*. Management Review Quarterly., DOI: 10.1007/s11301-022-00309-1

Okeyo, M. (2024). *Customer-centric data strategies to drive revenue growth*. eCommons@AKU. https://ecommons.aku.edu/theses_dissertations/2310/

Okumuş, M. S. (2022). *THE AUDIENCE STRUGGLE BETWEEN TELEVISION, CINEMA, AND OTT PLATFORMS: THE CASE OF NETFLIX AND DISNEY+ IN TURKEY. DergiPark*. Istanbul University., DOI: 10.7456/11301100/008

Paschmann, J. W., Bruno, H. A., Van Heerde, H. J., Völckner, F., & Klein, K. (2024). EXPRESS: Driving Mobile App User Engagement through Gamification. *JMR, Journal of Marketing Research*. Advance online publication. DOI: 10.1177/00222437241275927

Patil, D., Rane, N. L., & Rane, J. (2024). The future of customer loyalty: How ChatGPT and generative artificial intelligence are transforming customer engagement, personalization, and satisfaction. In *Science DIrect*. DOI: 10.70593/978-81-981367-8-7_2

Pizzo, A. D., Fornaciari, C. J., Coyle, P., Varzgani, N., & Pierce, M. (2024). Opening the metaverse: The next frontier in sport and leisure. *Leisure Sciences*, •••, 1–19. DOI: 10.1080/01490400.2024.2407023

Pratama, S. F. (2024). Temporal Patterns in user Conversions: Investigating the impact of ad scheduling in digital marketing. *Journal of Digital Market and Digital Currency.*, *1*(2), 165–182. DOI: 10.47738/jdmdc.v1i2.10

Putra, E. A. (2024, November 23). *The concept of Gharar in Purchasing Mobile Legends Virtual Items: A study of the clarity of objects and transactions*. Putra | Saqifah: Jurnal Hukum Ekonomi Syariah. https://journals.fasya.uinib.org/index.php/saqifah/article/view/638

Rabbi, M., Pfammatter, A., Zhang, M., Spring, B., & Choudhury, T. (2015). Automated Personalized feedback for physical activity and dietary behavior change with mobile phones: A randomized controlled trial on adults. *JMIR mHealth and uHealth*, *3*(2), e42. DOI: 10.2196/mhealth.4160 PMID: 25977197

Rabhan, B. (2013). Convert Every Click: Make More Money Online with Holistic Conversion Rate Optimization. In *Wiley Publishing eBooks*. https://dl.acm.org/citation.cfm?id=2566853

Radivojevic, K., Clark, N., Klempay, A., & Brenner, P. (2024). Defending novice user privacy: An evaluation of default web browser configurations. *Computers & Security*, *140*, 103784. DOI: 10.1016/j.cose.2024.103784

Ravichandran, N., Tewaraja, T., Rajasegaran, V., Kumar, S. S., Gunasekar, S. K. L., & Sindiramutty, S. R. (2024). Comprehensive Review Analysis and Countermeasures for Cybersecurity Threats: DDoS, Ransomware, and Trojan Horse Attacks. *Preprints.org*. DOI: 10.20944/preprints202409.1369.v1

Rege, I., & Dimitrova, N. (2024). *How does Ad Avoidance on phones affect Consumer Decision-Making?: An exploratory study on the effect thatad avoidance has on customer decision-making*. DIVA. https://www.diva-portal.org/smash/record.jsf?pid=diva2%3A1864770&dswid=-8184

Rendell, R., Pinheiro, M., Wang, B., McKay, F., Ewen, A., Carnegie, C., Tikomaidelana, E., Fattah, Z., & Hassett, L. (2024). Digital Apps to Improve Mobility in Adults with Neurological Conditions: A Health App-Focused Systematic Review. *Health Care*, *12*(9), 929. DOI: 10.3390/healthcare12090929 PMID: 38727486

Ringborg, L., & Ohlsson, I. (2024). *The influence of TikTok : Understanding the Influence of TikTok UGC on Gen Z's Impulsive Buying Behavior regarding Beauty Products*. DIVA. https://www.diva-portal.org/smash/record.jsf?pid=diva2%3A1862715&dswid=72

Rinpoche, D. K. (2024). *The free mind: Finding Clarity in a Digitally Distracted World*. Shambhala Publications.

Rodrik, D., & Sandhu, R. (2024). *Servicing Development: Productive upgrading of Labor-Absorbing services in developing economies*. DOI: 10.3386/w32738

Ros, R., Bjarnason, E., & Runeson, P. (2023). A theory of factors affecting continuous experimentation (FACE). *Empirical Software Engineering*, *29*(1), 21. Advance online publication. DOI: 10.1007/s10664-023-10358-z

Rutledge, P., Garner, M. S., & Collins, H. N. (2024). Exercising empowerment: Mediated fitness technology fostered social connections and personal growth during social isolation. *Psychology of Popular Media*. DOI: 10.1037/ppm0000517

Ryu, S. (2024). *Media and entertainment industry management: How to Integrate Business and Management with Creativity and Imagination*. Taylor & Francis. DOI: 10.4324/9781003271222

Sarangdhar, V., Awate, S., & Mudambi, R. (2024). Business model innovations in high-velocity environments. *Journal of Business Research*, *183*, 114864. DOI: 10.1016/j.jbusres.2024.114864

Sellgren, J. (2024). *Beyond the purchase : Exploring the dynamics of microtransactions among Swedish gamers*. DIVA. https://www.diva-portal.org/smash/record.jsf?pid=diva2%3A1867555&dswid=-7330

Seng, Y. J., Cen, T. Y., Raslan, M. H. B. M., Subramaniam, M. R., Xin, L. Y., Kin, S. J., Long, M. S., & Sindiramutty, S. R. (2024). In-Depth Analysis and Countermeasures for Ransomware Attacks: Case Studies and Recommendations. *Preprints.org*. DOI: 10.20944/preprints202408.2261.v1

Shah, I. A., Jhanjhi, N. Z., & Laraib, A. (2022). Cybersecurity and blockchain usage in contemporary business. In *Advances in information security, privacy, and ethics book series* (pp. 49–64). DOI: 10.4018/978-1-6684-5284-4.ch003

Sharma, R., Singh, A., Kavita, N., Jhanjhi, N. Z., Masud, M., Jaha, E. S., & Verma, S. (2021). Plant disease diagnosis and image classification using deep learning. *Computers, Materials & Continua/Computers. Materials & Continua (Print), 71*(2), 2125–2140. DOI: 10.32604/cmc.2022.020017

Sheil, A., Acar, G., Schraffenberger, H., Gellert, R., & Malone, D. (2024). *Staying at the Roach Motel: Cross-Country Analysis of Manipulative Subscription and Cancellation Flows*. ACM., DOI: 10.1145/3613904.3642881

Sheng, L. T., Azman, A. B., Khan, N. A., Sindiramutty, S. R., Singh, S., & Verma, S. (2024). Reinforcement learning-based traffic signal control. *2024 International Conference on Emerging Trends in Networks and Computer Communications (ETNCC)*, 792–798. DOI: 10.1109/ETNCC63262.2024.10767501

Shu-Chen, C. (2023). From freemium to premium app services: The Expectation confirmation model and Two-Factor Theory. *International Journal of Marketing Communication and New Media, 11*(21). Advance online publication. DOI: 10.54663/2182-9306.2023.v11.n21.112-131

Shukla, R. S., Ghuse, E. A., Diwan, T., Tembhurne, J. V., & Sahare, P. (2024). GA-MPG: Efficient genetic algorithm for improvised mobile plan generation. *Journal of Ambient Intelligence and Humanized Computing, 15*(10), 3675–3691. DOI: 10.1007/s12652-024-04846-3

Sindiramutty, S. R. (2024). Autonomous Threat Hunting: a future paradigm for AI-Driven Threat intelligence. *arXiv (Cornell University)*. https://doi.org//arxiv.2401.00286DOI: 10.48550

Sindiramutty, S. R., Prabagaran, K. R. V., Akbar, R., Hussain, M., & Malik, N. A. (2024). Overview of Generative AI techniques for cybersecurity. In *Advances in information security, privacy, and ethics book series* (pp. 1–52). DOI: 10.4018/979-8-3693-5415-5.ch001

Sindiramutty, S. R., Prabagaran, K. R. V., Jhanjhi, N. Z., Ghazanfar, M. A., Malik, N. A., & Soomro, T. R. (2024). Security Considerations in Generative AI for web Applications. In *Advances in information security, privacy, and ethics book series* (pp. 281–332). DOI: 10.4018/979-8-3693-5415-5.ch009

Sindiramutty, S. R., Prabagaran, K. R. V., Jhanjhi, N. Z., Murugesan, R. K., Brohi, S. N., & Masud, M. (2024). Generative AI in network security and intrusion detection. In *Advances in information security, privacy, and ethics book series* (pp. 77–124). DOI: 10.4018/979-8-3693-5415-5.ch003

Sindiramutty, S. R., Prabagaran, K. R. V., Jhanjhi, N. Z., Murugesan, R. K., Brohi, S. N., & Wei, G. W. (2024). Generative AI for threat intelligence and information sharing. In *Advances in digital crime, forensics, and cyber terrorism book series* (pp. 191–234). DOI: 10.4018/979-8-3693-8944-7.ch006

Sindiramutty, S. R., Prabagaran, K. R. V., Jhanjhi, N. Z., Murugesan, R. K., Malik, N. A., & Hussain, M. (2024). Ethics and transparency in secure web model generation. In *Advances in information security, privacy, and ethics book series* (pp. 411–464). DOI: 10.4018/979-8-3693-5415-5.ch012

Singhal, V., Jain, S. S., Anand, D., Singh, A., Verma, S., Kavita, N., Rodrigues, J. J. P. C., Jhanjhi, N. Z., Ghosh, U., Jo, O., & Iwendi, C. (2020). Artificial Intelligence Enabled Road Vehicle-Train Collision Risk Assessment Framework for Unmanned railway level crossings. *IEEE Access : Practical Innovations, Open Solutions*, *8*, 113790–113806. DOI: 10.1109/ACCESS.2020.3002416

Suhod, M. S. M., Ngah, A. H., & Rahi, S. (2024). Unravelling the continue of subscribing Spotify Premium among university students: The extended UTAUT 2 model. *International Journal of Business Information Systems*, *46*(2), 217–237. DOI: 10.1504/IJBIS.2024.139908

Tirole, J. (2024). Competition and industrial policy in the 21st century. *Oxford Open Economics*, *3*(Supplement_1), i983–i1001. DOI: 10.1093/ooec/odad080

Varadarajan, M. N., & Seeni, S. K. (2024). INNOVATIVE DIGITAL OWNERSHIP AND COLLECTIBLES VIA PROOF OF STAKE (POS) AND NON-FUNGIBLE TOKENS (NFTS). *International Journal of Advances in Signal and Image Sciences*, *10*(1), 22–34. DOI: 10.29284/IJASIS.10.1.2024.22-34

Waheed, A., Seegolam, B., Jowaheer, M. F., Sze, C. L. X., Hua, E. T. F., & Sindiramutty, S. R. (2024). Zero-Day Exploits in Cybersecurity: Case Studies and Countermeasure. *preprints.org*. DOI: 10.20944/preprints202407.2338.v1

Weiqi, X., Hooi, S. T. C., Sindiramutty, S. R. L., Asirvatham, D. L., Kumar, D., & Verma, S. (2024). Surface Anomaly Detection Using Machine Learning Technique. *2024 International Conference on Emerging Trends in Networks and Computer Communications (ETNCC)*, 1–7. DOI: 10.1109/ETNCC63262.2024.10767562

Wen, B. O. T., Syahriza, N., Xian, N. C. W., Wei, N. G., Shen, T. Z., Hin, Y. Z., Sindiramutty, S. R., & Nicole, T. Y. F. (2023). Detecting cyber threats with a Graph-Based NIDPS. In *Advances in logistics, operations, and management science book series* (pp. 36–74). DOI: 10.4018/978-1-6684-7625-3.ch002

What is monetization? | Adjust. (2024). https://www.adjust.com/glossary/app-monetization/

Yan, C., Yang, Z., & Dai, X. (2023). Subscriptions versus one-off purchases: The impact of fee-charging models on consumer privacy concerns. *International Journal of Bank Marketing, 41*(4), 835–859. DOI: 10.1108/IJBM-06-2022-0271

Ying, X., Murugesan, R. K., Sindiramutty, S. R., Wei, G. W., Balakrishnan, S., Kumar, D., & Verma, S. (2024). Scene Text Recognition using Deep Learning Techniques. *2024 International Conference on Emerging Trends in Networks and Computer Communications (ETNCC)*, 1–9. DOI: 10.1109/ETNCC63262.2024.10767484

Youn, K., Pittman, M., & Haley, E. (2023). Using the FCB grid to understand privacy concerns in social media advertising. *International Journal of Advertising, 43*(3), 405–433. DOI: 10.1080/02650487.2023.2242020

Zhang, H., & Sun, Q. (2023). The transformation mechanism of fitness clubs: Pricing of joint fitness courses by online platforms and well-known coaches. *Journal of Retailing and Consumer Services, 76*, 103539. DOI: 10.1016/j.jretconser.2023.103539

Zhang, K., Wang, Q., Qiu, L., & Wang, N. (2024). Unveiling the cost of free: How an Ad-Sponsored model affects serialized digital content creation. *Information Systems Research.* Advance online publication. DOI: 10.1287/isre.2023.0263

Zhang, X., Kamarudin, S., Tang, Q., & Cai, Y. (2024). Within the lens of "Keep": The convergence of consumer context and brand development in Chinese fitness apps. *International Journal of Academic Research in Business & Social Sciences, 14*(1). Advance online publication. DOI: 10.6007/IJARBSS/v14-i1/20004

Zhou, Z., Lin, L., Wang, H., Zhou, X., Wei, G., & Wang, S. (2024, May). A Cross Domain Method for Customer Lifetime Value Prediction in Supply Chain Platform. In *Proceedings of the ACM Web Conference 2024* (pp. 4037-4046).

Chapter 7
Detecting and Analyzing Malware in Android Mobile Devices Through Network Traffic Monitoring

Khalid Hafiz Mir
https://orcid.org/0009-0007-2610-3779
Lovely Professional University, India

Anzah Bashir
Lovely Professional University, India

Isha Batra
Lovely Professional University, India

ABSTRACT

Attacks by malevolent software developers have increased in tandem with the growing popularity of cell phones. This study examines several techniques these assailants employ and recommends more effective ones. Android phones are especially susceptible to these attacks since network traffic analysis isn't taken into account sufficiently in today's detection techniques. In this article, a rule-based classifier—a cutting-edge technique—that can reliably identify more than 90% of hazardous traffic is introduced. Malicious apps are becoming more and more prevalent due to the expanding black market and the increasing number of applications that are accessible. To identify questionable activity, the suggested system automatically collects and examines applications from reputable retailers or unofficial marketplaces. The

DOI: 10.4018/979-8-3693-6925-8.ch007

research provides a clear picture of how the platform has changed over time and how well various detection techniques work. It also compares several techniques for Android malware detection.

1. INTRODUCTION

Due to their widespread use, cell phones are also a popular target for online criminals. Smartphones are increasingly being utilized for personal data storage and corporate transactions, which makes them susceptible to malware assaults. Because consumers frequently lack the time to carefully consider every program they download from websites or app stores, malicious developers find it simpler to target mobile devices. Malware may take many different forms, such as trojans, spyware, viruses, and phishing programs. Phishing apps imitate genuine apps, but their main goal is to get login passwords and sensitive data for fraudulent financial transactions(Arora et al., 2014). While viruses attach themselves to executable files and proliferate, Trojans pose as trustworthy applications to obtain illegal access to machines. Google's Android has surpassed previous smartphone systems like Symbian to become the most widely used. The proliferation of feature-rich apps in online marketplaces is partly responsible for the exponential increase in Android malware. Although legitimate software stores like Google Play have certain anti-malware features, their main security feature is permissions, which isn't always reliable(Zaheer et al., 2022). The problem is made worse by third-party app marketplaces, which permit software deployment without strict security controls. Attackers use a variety of strategies, such as update assaults, drive-by downloads, and repackaging, to target cellphones. Malware makers, in particular, frequently repackage programs, including dangerous code, into them before redistributing them through official and unofficial shops. Because certain Android viruses can collect personal data and send it to distant locations, they are very harmful. It is critical to address the threat posed by Android malware.

This study presents a system for remotely identifying Android malware that frequently connects to a remote server to transmit sensitive information or is controlled by a server. More reliable methods for identifying Android malware may be created by applying rule-based classifier development and network traffic feature analysis(Wang et al., 2019). App usage is on the rise as a result of the many activities that smartphone users partake in, including messaging, calling, social networking, and online surfing. The unauthorized third-party markets known as Black markets and legitimate app stores like Google Play and the App Store have both experienced explosive development. However, hackers may easily spread dangerous software in these stores because there are no code reviews or security checks in place.

Trojan-SMS. AndroidOS, Geinimi, Android/Spyware, PJApps.a, and Gingerbreak are some of the most common forms of Android malware. The automated gathering, examination, and reporting of apps is part of the suggested architecture for identifying harmful activity in Android apps. It has capabilities including dynamic analysis based on patterns of dangerous behavior, automated software gathering from official and unofficial stores, and static analysis for extracting unsafe strings and APIs(Casolare et al., 2021). Because Android is so popular for accessing digital services on a wide range of devices, users and developers need to take precautions to ensure security(Ashraf et al., 2023). The current focus of attempts to tackle Android malware is on creating sophisticated detection techniques that frequently make use of machine learning models. Nonetheless, many anti-malware providers continue to use signature-based techniques, which are easily beaten with a few straightforward code changes.

2. ARCHITECTURE FOR MALWARE IDENTIFICATION IN ANDROID-BASED MOBILE DEVICES VIA NETWORK TRAFFIC ANALYSIS:

To create an architecture for Malware Identification in Android-Based Mobile Devices via Network Traffic Analysis, the system could be structured as follows:

Figure 1. Architecture for Malware Identification in Android-Based Mobile Devices via Network Traffic Analysis

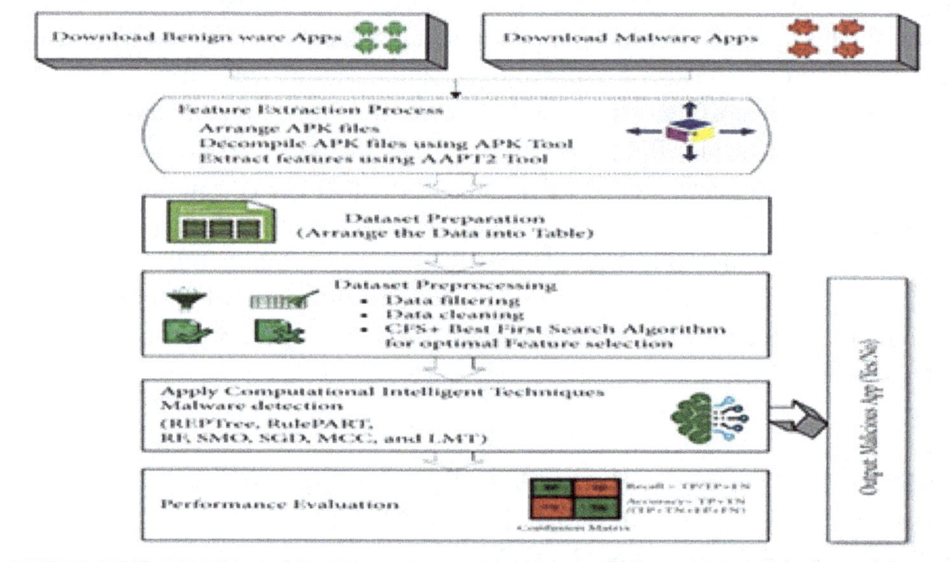

2.1 Data Collection Layer

- Mobile Devices: Android devices serve as the primary source of network traffic. These devices can run both benign and potentially malicious applications.
- Network Traffic Capturing Tools: Tools such as tcpdump or Wireshark capture network packets (HTTP, DNS, SSL/TLS, etc.) from Android devices. This can also be implemented using Android's VPN API to intercept traffic.
- Cloud or Local Proxy Servers: Devices send traffic through a proxy to monitor and capture the data in a centralized way(Al-Quayed et al., 2024).

2.2 Preprocessing Layer

- Packet Parsing Module: Extracts relevant features (e.g., source/destination IP addresses, ports, protocols, and payload size) from raw network packets.
- Traffic Aggregation: Organizes the network traffic into flows (sequence of packets between source and destination) for further analysis(Abbasi et al., 2021).

- Feature Extraction: Key features such as connection duration, packet size, header information, and anomalies in protocols are extracted for each traffic flow.

2.3 Detection/Analysis Layer

- Machine Learning Models:
  - A supervised or unsupervised ML model (e.g., decision trees, random forest, SVM, neural networks) is trained to detect malware based on the network traffic features.
  - Models are continuously updated as new types of malware emerge.
- Anomaly Detection: Behavior-based methods can detect deviations from normal traffic patterns, flagging them as suspicious.
- Signature-based Detection: Uses known malware patterns to identify known threats.
- Heuristic/Behavioral Analysis: Observes unusual behaviors like frequent communication with command-and-control servers, large data exfiltration, or abnormal DNS requests.

2.4 Storage Layer

- Data Storage: Stores captured traffic data for further offline analysis.
- Malware Signatures Database: A continuously updated database with known malware patterns, signatures, and characteristics.
- Traffic Patterns Repository: Stores patterns of normal vs. anomalous traffic behaviors for baseline comparison.

2.5 Alert and Reporting Layer

- Real-time Alerts: When malware activity is detected, alerts are sent to system administrators for immediate action (e.g., blocking malicious IP addresses or applications)(Aldughayfiq et al., 2023).
- Reporting Dashboard: Visualizes traffic trends, identified malware, and potential risks to provide an overview of the system's health and security status.
- Logs and Reports: Detailed reports of malware detections, traffic logs, and suspicious behaviors for auditing and forensic analysis.

2.6 Response Layer

- Incident Response: Automatically isolates or blocks malicious network activities when malware is detected.
- Device Remediation: Sends commands to affected Android devices to quarantine or uninstall malicious apps.
- Updates and Patches: Pushes updates to malware signature databases and machine learning models as new malware emerges.

3. METHODS FOR DETECTING MALWARE

Static, dynamic, and hybrid approaches are the three categories of mobile malware detection strategies; each has advantages and disadvantages of its own(Tong & Yan, 2017).

3.1 Method for Static Analysis Detection

Programs are analyzed statically—that is, without being run. Reverse engineering methods and techniques are used to recreate the application's source code and algorithms during static analysis. Debuggers, disassemblers, and program analyzers may all be used for static analysis. Several static techniques consist of:

3.2 Signature-Based Detection Technique

Also referred to as fingerprinting or pattern matching, this technique entails malware authors inserting a signature into the application code to specifically identify certain malware. It is necessary to search the code for pre-defined signatures to detect them(M. & Sethuraman, 2023).

3.3 Heuristic Method for Detection

Alternatively referred to as a proactive approach, this technique looks for commands or instructions that are missing from the application software instead of looking for certain signatures. This facilitates the detection of novel malware variants. Heuristic analysis techniques include:

- File-based heuristic analysis: Carefully examine the file in light of its objectives, content, and purpose. We identify malicious files that include instructions to destroy or damage other files.

- Weight-based heuristic analysis: Gives each application a risk score, marking those that go beyond a set threshold as possibly harboring harmful code(Verma et al., 2023).
- Heuristic analysis based on rules: identifies and compares the application's rules to pre-established ones. Applications with intricate rules are marked as possibly harmful.
- Generic signature analysis uses antivirus signatures to find novel varieties of malware, allowing for the detection of variants with distinct characteristics within the same cohort.

3.4 Static Analysis Tools

These instruments are used in an initial examination to identify potential security risks in dubious applications:

- IDA Pro classifies apps according to their harmful actions by extracting system calls that the application makes and doing anomaly detection.

Figure 2. IDA process

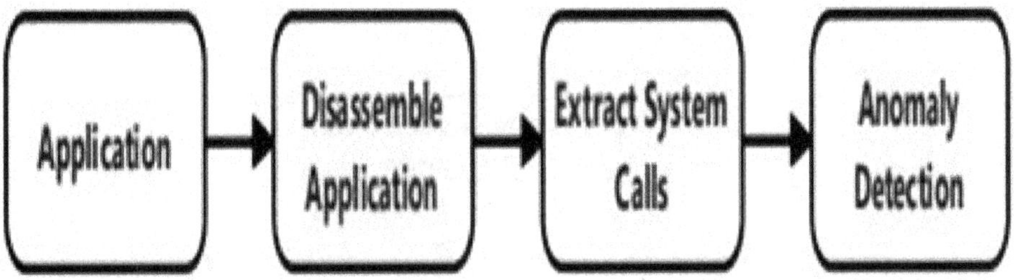

- PiOS: Constructs a control flow chart from application binaries to assess an application's network transmission capabilities and access sensitivities.
- UPX: A powerful executable packer with a quick decompression rate and good compression ratio.
- ProcDump: Tracks CPU spikes in programs and produces crash dumps to identify the source of the surge.

3.5 Method for Dynamic Analysis Detection

Analysis of an application's behavior during execution is known as dynamic analysis. Function calls, data streams, function parameter analysis, and execution path tracing are all included(Shijo & Salim, 2015). Applications are usually executed in a controlled environment for dynamic analysis using a virtual machine, often known as a sandbox. Detection techniques consist of:

3.6 Anomaly-Based Detection

Suggested tools dynamically track the activity of applications and produce logs that are sent to a distant server for examination. Using device state monitoring, Andromly, for instance, uses machine learning techniques to differentiate between benign and malicious apps.

3.7 Emulation-Based Detection

DroidScope and similar platforms use virtual machine introspection to keep an eye out for harmful activity across the operating system. One such technology is the Android Application Sandbox.

Figure 3. Methods for detecting malware

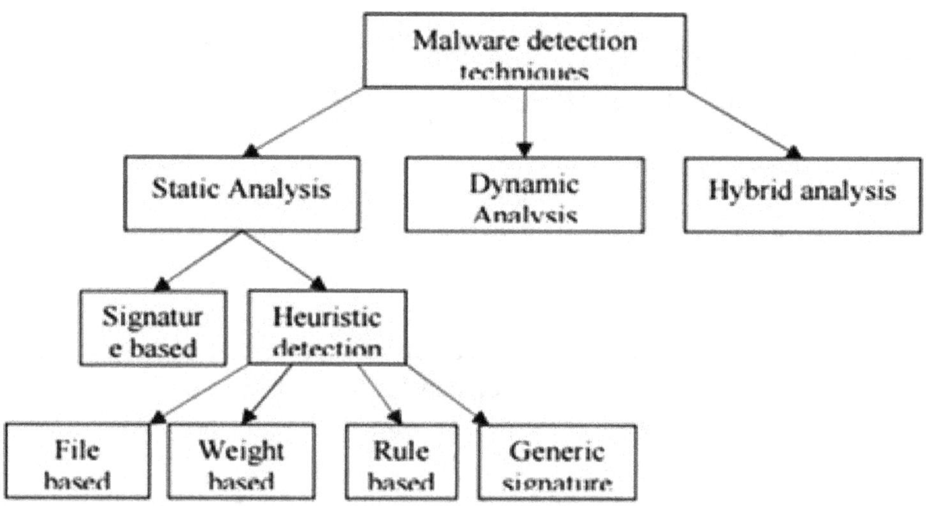

The suggested method involves many implementation phases, such as data collection, analysis, detection, feature selection, and classifier selection. Let's go over each step in more detail:

• *Selection of Features*

In network security, a variety of traffic characteristics are used to identify malicious behavior. Table 1 summarizes the characteristics selected for the proposed study based on the behavior of several malware variants.

Table 1. LIST OF NETWORK TRAFFIC FEATURES

Notation	*Traffic Feature*
Apps	The average size of the Packet
Adf	The flow's average duration
Tps	Time Interval Between Packets Sent
Iob	The ratio of Received to Sent Bytes
Iop	Incoming to Outgoing Packet Ratio
Bsf	The average amount of data sent for each flow
Tpr	The interval of time between received packets
TrC	Traffic Content (percentage of TCP)
Psf	The mean quantity of packets sent for each flow
Bss	Average Bits Transferred in a Second
Pss	Average packets Transferred in a Second
Brf	The mean quantity of bytes obtained for each flow
Prf	The mean quantity of packets obtained for each flow
Brs	The average amount of data received in a second
Prs	The average packets received in a second
CtD	Number of connections to destination IPs as a ratio

• Choosing a Classifier

The literature has presented a wide range of machine learning classification methods for categorizing data values according to attribute sets. Because rule-based classifiers are very expressive and resemble decision trees, they were chosen for this particular project. Rule-based classifiers may quickly categorize new cases and are simple to understand(Kliegr & Izquierdo, 2023).

Samples of Malware

The Android Malware Genome Project at North Carolina State University provided the Android malware samples utilized in the studies(Guido et al., 2013). Malware that can be remotely operated or that sends data to a distant server is the main focus of this study. Just 13 of the 27 malware families that were detected in the research reacted in the testing. Table II contains a list of these 13 respondent samples. It's possible that other malware families aren't responding because they can't connect to the downed remote server.

Information Gathering

Using an Android emulator that is set up on a host computer with a public IP address is how data gathering is done. The emulator isolates the host machine's network interface from itself by using a virtual router with a fixed IP address of 10.0.2.15. For around fifteen hours, one sample of every virus listed in Table II is run on the Android emulator. Every malicious program appears as an infected Android app and establishes a background connection to a remote server(Qamar et al., 2019). The tcpdump command and the adb tool (Android Debug Bridge) are used on the host system to collect the network traffic produced by these applications. Then, Wireshark is used to examine all of the traffic that was collected. Additionally, the same tcpdump setup used to capture malware traffic is also used to record three examples of innocuous mobile data, including web surfing, an online dictionary app, and an online newspaper app.

Table 2. A LIST OF ANDROID MALWARE AND THE PERSONALITIES OF EACH

Malwares	Characteristics of Malwares
AnserverBot	Has the ability to load dynamic code and obfuscate deep code to prevent detection
BaseBridge	Contains code capable of killing any security application and sending SMS messages
BgServ	After gathering information such as the IMEI, phone number, and system version, it downloads data to the server.
DroidDream	It uploads the device model, SDK version, IMEI, and IMSI to the server.
DroidDreamLight	A DroidDream variant with the ability to switch its C&C server
DroidKungFu1	It sends the Android OS version, phone model, and IMEI to the server.
DroidKungFu2	Variant of DroidKungFu1 that has two C&C domains added to it
DroidKungFu3	Obscures all native binaries connected to malware and obscures external C&C server URLs via encryption
GoldDream	It uploads the distinct subscriber ID and IMEI number to the server.
jSMSHider	It sends the phone number and IMSI to the server.

continued on following page

Table 2. Continued

Malwares	Characteristics of Malwares
Pjapps	It uploads IMEI, Device ID, Line number, Subscriber ID and SIM Serial Number to the server
RogueLemon	Dependent on the instruction received from the server, it is capable of sending SMS without the user's awareness.

Using Wireshark, the traffic analysis process started by transforming the collected packets into and UDP flow discussions. To extract the 16 attributes listed in Table I from these chats, a shell script was created for each of them(Sprengers & van Haaster, 2016). The majority of malware linked to external servers on a regular basis, according to observations. Plankton transmitted HTTP POST packets every 4 seconds, BaseBridge received instructions from the server every 30 seconds, and AnserverBot connected to the server every two hours.

Figure 4. Malware Traffic Capturing Experimental Setup

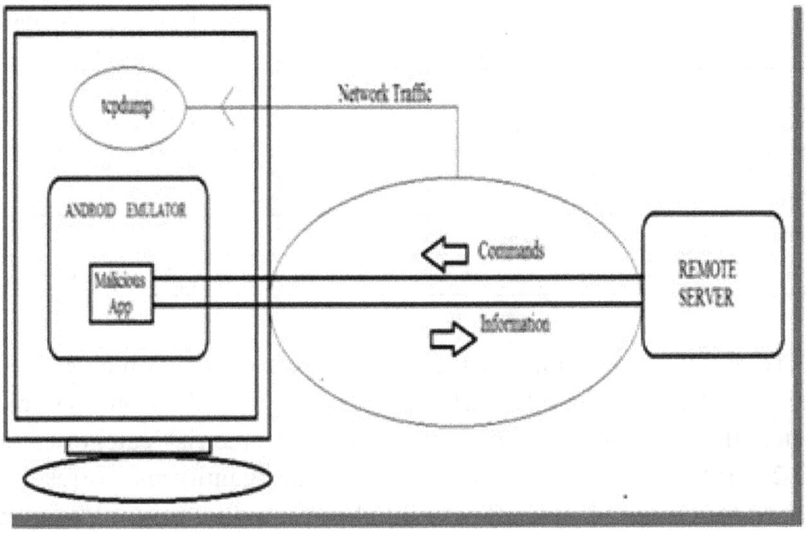

The examination of traffic characteristics is summarized in Table III, which also reveals seven factors that set malware traffic apart from regular mobile traffic. Malware was found to have an average packet size of between 91 and 352 bytes, which was less than typical mobile traffic (which ranges from 600 to 900 bytes) (Zarpelão et al., 2017). This behavior is similar to that of computer-based botnets. In addition, characteristics like the Average Number of Bytes Received per Second and the Ratio of Incoming to Outgoing Bytes were much greater for legitimate traffic than for malicious traffic(Tsikerdekis et al., 2021).

Table 3. ANALYSIS OF TRAFFIC FEATURES

Traffic Features	Malware Traffic	Normal Traffic
Packet Size Average (in bytes)Incoming & Outgoing Bit Ratio	91-342	699-990
The average amount of data received in a second	0.1-2.8	7-8
The mean quantity of packets sent for each flow	29-1300	14000-16000
The mean quantity of packets obtained for each flow	5-8	20-40
The average amount of data sent for each flow	1-7	26-65
The typical quantity of information obtained for every flow	500-1700	2400-3700
The average number of packets transmitted in a second	500-4900	30000-88000
Average packet count received in a second	0.1-28	14-17
Average Bits Transferred in a Second	0.1-28	15-18
The interval of time between sent packets	11-5000	17-2000
The interval of time between received packets	0.1-116	0.5-0.8
The flow's average duration, measured in seconds	0.1-139	0.5-0.9
Inflow to Outflow Packet Ratio per Flow	0.6-700	7-11
Ratio of Incoming to Outgoing Packets per flow	0.8-1	0.9-1
Number of connections to destination IPs as a ratio	1-60	3-4
Traffic Content (percentage of TCP)	More than 90%	More than 90%

To assess the accuracy of the classifier, a total of 43 malware samples and 5 samples of regular mobile traffic were used. Depending on how many attributes match, the suggested work divides traffic into three risk categories:(Smmarwar et al., 2024) low, medium, and high. Based on the quantity and corresponding proportion of matching characteristics, Table V defines these risk categories Since all of the chosen attributes helped to differentiate malware traffic from legitimate traffic, they were all given the same priority. For example, if any two characteristics in the data matched the spectrum of malware, the algorithm rated the traffic as low-risk. Using an Android emulator, all malware samples were allowed to operate for a full day, and their network traffic was captured using the tcpdump option. As a result, 43 days passed before the traffic of 43 malware samples was captured.

Table 4. DESCRIPTION OF RISK LEVEL

Risk-Level	Number of Features (out of 7) Matched with Malware Traffic	Percentage of features that have malware traffic matched to them
Low	0-2	< 30%
Medium	3-5	30-70%
High	6-7	>70%

4. CHALLENGES IN MALWARE DETECTION ON ANDROID DEVICES

Malicious apps created to inflict harm on Android users and devices are known as Android malware. Credential theft, data leaks, inserting malicious code into trustworthy programs, deleting or encrypting files, and changing device settings are a few examples of these actions. Android has been a top target for cybercriminals because it is the most widely used operating system on smartphones. Below are some key challenges, particularly related to evasion techniques, resource limitations, and encrypted traffic(Yousaf et al., 2017).

- Mobile Malware Evasion Techniques: To evade detection, malware employs stealthy network behavior, encryption, domain fluxing, and obfuscation. Malware conceals itself from detection via packing payloads, obfuscating code, employing domain fluxing, encrypting communication, and imitating authentic network activity.
- Limitations on Device Resources: Real-time malware detection on Android smartphones is hampered by low CPU, memory, and battery life. Resource-intensive security solutions cannot be used on Android smartphones due to their limited CPU, memory, and battery life, and network unpredictability makes regular traffic analysis and anomaly detection more difficult.
- The function of encrypted traffic is to mask malicious activity, which makes network traffic analysis challenging. Examples of encrypted protocols are HTTPS and E2EE. It is challenging to examine data packets for harmful information or commands because of the widespread usage of end-to-end encryption (E2EE), encrypted C2 communication, and HTTPS, which hide malicious activity from network traffic analysis tools.

5. DATA COLLECTION FOR TRAFFIC ANALYSIS ON ANDROID DEVICES:

Important techniques for gathering network traffic on Android smartphones consist of:

- Virtual Network Interface (VPN)-Based Traffic Collection: This method collects traffic without requiring the device to be root-accessed, although it may overlook lower-level communications and cause privacy issues.
- Rooted Devices: Offer complete system access, enabling in-depth traffic analysis, but are less prevalent in practical applications and come with security hazards.
- Third-Party Applications and Proxies: Programs such as Wireshark and Charles Proxy use a local proxy to intercept traffic, but they might not be able to collect all of it, particularly from programs that use encrypted communication.

Trade-offs exist between each method's compatibility, security, and data.

5.1 Privacy Concerns and Ethical Considerations

- User Consent and Transparency: It's essential to obtain informed consent from users before collecting network traffic data, explaining what will be collected, why, and how it will be used. This ensures ethical data collection practices and user awareness.
- Sensitive Data Protection: Network traffic may include personal information like usernames, passwords, or messages. To prevent privacy breaches, anonymization and encryption must be applied, and strict access controls should be enforced during analysis.
- Balancing Detection and Privacy: Detailed traffic analysis improves malware detection but can expose personal data. Ethical challenges arise with end-to-end encryption (E2EE), as breaking encryption for security may infringe on user privacy, posing a dilemma between privacy and security.
- Legal Compliance: Traffic analysis must comply with regulations such as GDPR, requiring clear user consent and data protection to avoid legal consequences. Organizations must align their practices with legal standards to protect users' rights.

5.2 Dealing with Large-Scale Network Traffic Datasets

Handling large-scale network traffic data from Android devices for malware detection requires overcoming several challenges related to storage, processing, and real-time analysis:

- Efficient Storage and Compression: Massive amounts of traffic data necessitate efficient compression and structured storage systems (e.g., Hadoop, HDFS, or cloud storage) to manage the data without exceeding storage limits. Partitioning data by time intervals or features like IP addresses can help with querying and managing the data effectively.
- Processing Large Datasets: Distributed computing frameworks such as Apache Spark or MapReduce enable parallel processing of large datasets. Techniques like filtering and feature extraction (e.g., packet size, timing) streamline analysis by focusing on potentially malicious traffic.
- Real-Time Analysis: Stream processing platforms (e.g., Apache Kafka, Flink) allow real-time analysis of traffic as it's generated, facilitating timely detection of suspicious activity. Machine learning models, especially unsupervised learning, help detect anomalies in real-time, but these require robust resources for model training and updates.
- Labeling and Ground Truth: Properly labeling network traffic as malicious or benign remains a challenge due to its dynamic nature. Combining manual analysis, heuristics, and external threat intelligence helps in accurately identifying malware-related patterns.

In summary, effective traffic analysis for Android malware detection requires the right mix of storage strategies, processing frameworks, real-time monitoring, and labeling techniques, while ensuring user privacy and legal compliance.

6. FUTURE DIRECTIONS AND EMERGING THREATS IN ANDROID MALWARE DETECTION:

6.1 Trends in Android Malware Development and Distribution

Android malware continues to evolve, with developers employing new strategies to bypass security measures. Notable trends include:

6.2 Modular malware: Malware is becoming more modular, allowing different components to be downloaded separately, making detection harder as the entire payload isn't present at once.
6.3 Malware-as-a-Service (MaaS): Similar to traditional cybercrime, Android malware is being commoditized, where developers sell or lease malware to other attackers, increasing the frequency and sophistication of attacks.
6.4 Targeting financial apps: Attackers increasingly focus on banking and cryptocurrency apps, deploying trojans to steal credentials, initiate fraudulent transactions, or intercept two-factor authentication (2FA) codes.
6.5 Use of zero-day vulnerabilities: Exploiting unpatched vulnerabilities in the Android ecosystem to gain unauthorized access before manufacturers can release security patches.
6.6 Role of 5G and IoT in Expanding the Attack Surface:

The rollout of 5G and the proliferation of IoT devices significantly increases the potential attack surface for Android malware:

6.7 5G Networks: The ultra-fast, low-latency 5G networks enable a greater number of connected devices and data-intensive applications, which provides more opportunities for malware to propagate. The higher speeds also allow attackers to launch more sophisticated attacks in real time(Chowdhury, n.d.).
6.8 IoT Integration: As Android is widely used in IoT devices, such as smart TVs, home automation systems, and wearable tech, these devices become vulnerable entry points. Many IoT devices lack robust security measures, making them easy targets for malware that can spread across networks.
6.9 Increased device mobility: With 5G supporting constant connectivity across various environments (e.g., smart cities, autonomous vehicles), malware can exploit the seamless transitions between networks and devices, complicating detection.

7. ADVANCED DETECTION TECHNIQUES USING AI AND MACHINE LEARNING:

AI and machine learning are at the forefront of evolving malware detection capabilities. Key innovations include:

7.1 Behavioral analysis: Machine learning models can be trained to detect anomalies in device behavior, such as unusual patterns in network traffic or system resource usage, flagging potential malware before traditional signature-based methods could detect it.
7.2 Unsupervised learning: This approach allows models to detect previously unknown threats by identifying deviations from normal patterns in large datasets without relying on labeled training data.
7.3 Automated malware classification: AI-based systems can classify malware types based on patterns in their communication behavior, payload, and system interactions. This allows for faster categorization and response to emerging threats.
7.4 Federated learning: In a privacy-preserving manner, federated learning allows AI models to be trained across many devices without collecting the data centrally. This helps improve detection accuracy without compromising user privacy

8. CONCLUSION

This study proceeds a traffic analysis to identify malware on Android smartphones by examining their network traffic features. This study reflects the pioneering effort in Android malware identification based on network traffic analysis. It specifically targets malware that creates network traffic by talking to remote sites. Although it's not generally applicable to all kinds of malware, the method gives satisfactory accuracy in the results. Future study will increase by selecting features, improving detection accuracy, and analyzing bigger datasets.

REFERENCES

Abbasi, M., Shahraki, A., & Taherkordi, A. (2021). Deep Learning for Network Traffic Monitoring and Analysis (NTMA): A Survey. *Computer Communications*, *170*, 19–41. DOI: 10.1016/j.comcom.2021.01.021

Al-Quayed, F., Javed, D., Jhanjhi, N. Z., Humayun, M., & Alnusairi, T. S. (2024). A Hybrid Transformer-Based Model for Optimizing Fake News Detection. *IEEE Access : Practical Innovations, Open Solutions*, *12*, 160822–160834. DOI: 10.1109/ACCESS.2024.3476432

Aldughayfiq, B., Ashfaq, F., Jhanjhi, N. Z., & Humayun, M. (2023). YOLO-Based Deep Learning Model for Pressure Ulcer Detection and Classification. *Health Care*, *11*(9), 1222. DOI: 10.3390/healthcare11091222 PMID: 37174764

Arora, A., Garg, S., & Peddoju, S. K. (2014). Malware Detection Using Network Traffic Analysis in Android Based Mobile Devices. *2014 Eighth International Conference on Next Generation Mobile Apps, Services and Technologies*, 66–71. DOI: 10.1109/NGMAST.2014.57

Ashraf, H., Khan, F., Ihsan, U., Al-Quayed, F., Jhanjhi, N. Z., & Humayun, M. (2023). MABPD: Mobile Agent-Based Prevention and Black Hole Attack Detection in Wireless Sensor Networks. *2023 International Conference on Business Analytics for Technology and Security (ICBATS)*, 1–11. DOI: 10.1109/ICBATS57792.2023.10111277

Casolare, R., De Dominicis, C., Iadarola, G., Martinelli, F., Mercaldo, F., & Santone, A. (2021). Dynamic mobile malware detection through system call-based image representation. *Journal of Wireless Mobile Networks, Ubiquitous Computing and Dependable Applications*, *12*(1), 44–63. DOI: 10.22667/JOWUA.2021.03.31.044

Chowdhury, A. T. (n.d.). *The Impact Of 5G Network On Industry 4.0*. DOI: 10.13140/RG.2.2.25308.16001

Guido, M., Ondricek, J., Grover, J., Wilburn, D., Nguyen, T., & Hunt, A. (2013). Automated identification of installed malicious Android applications. *Digital Investigation*, *10*, S96–S104. DOI: 10.1016/j.diin.2013.06.011

Kliegr, T., & Izquierdo, E. (2023). QCBA: Improving rule classifiers learned from quantitative data by recovering information lost by discretisation. *Applied Intelligence*, *53*(18), 20797–20827. DOI: 10.1007/s10489-022-04370-x

Qamar, A., Karim, A., & Chang, V. (2019). Mobile malware attacks: Review, taxonomy & future directions. *Future Generation Computer Systems*, *97*, 887–909. DOI: 10.1016/j.future.2019.03.007

Shijo, P. V., & Salim, A. (2015). Integrated Static and Dynamic Analysis for Malware Detection. *Procedia Computer Science*, *46*, 804–811. DOI: 10.1016/j.procs.2015.02.149

Smmarwar, S. K., Gupta, G. P., & Kumar, S. (2024). Android malware detection and identification frameworks by leveraging the machine and deep learning techniques: A comprehensive review. *Telematics and Informatics Reports*, *14*, 100130. DOI: 10.1016/j.teler.2024.100130

Sprengers, M., & van Haaster, J. (2016). Organization of #operations. In *Cyber Guerilla* (pp. 41–109). Elsevier. DOI: 10.1016/B978-0-12-805197-9.00003-6

Tong, F., & Yan, Z. (2017). A hybrid approach of mobile malware detection in Android. *Journal of Parallel and Distributed Computing*, *103*, 22–31. DOI: 10.1016/j.jpdc.2016.10.012

Tsikerdekis, M., Waldron, S., & Emanuelson, A. (2021). Network Anomaly Detection Using Exponential Random Graph Models and Autoregressive Moving Average. *IEEE Access : Practical Innovations, Open Solutions*, *9*, 134530–134542. DOI: 10.1109/ACCESS.2021.3116575

Verma, V., Malik, A., & Batra, I. (2023). Assessing the Impact and Performance of Malware Detection Systems. *2023 10th IEEE Uttar Pradesh Section International Conference on Electrical, Electronics and Computer Engineering (UPCON)*, 106–109. DOI: 10.1109/UPCON59197.2023.10434443

Wang, S., Chen, Z., Yan, Q., Yang, B., Peng, L., & Jia, Z. (2019). A mobile malware detection method using behavior features in network traffic. *Journal of Network and Computer Applications*, *133*, 15–25. DOI: 10.1016/j.jnca.2018.12.014

Yousaf, F. Z., Bredel, M., Schaller, S., & Schneider, F. (2017). NFV and SDN—Key Technology Enablers for 5G Networks. *IEEE Journal on Selected Areas in Communications*, *35*(11), 2468–2478. DOI: 10.1109/JSAC.2017.2760418

Zaheer, A., Tahir, S., Humayun, M., Almufareh, M. F., & Jhanjhi, N. Z. (2022). A novel Machine learning technique for fake smart watches advertisement detection. *2022 14th International Conference on Mathematics, Actuarial Science, Computer Science and Statistics (MACS)*, 1–5. DOI: 10.1109/MACS56771.2022.10023151

Zarpelão, B. B., Miani, R. S., Kawakani, C. T., & de Alvarenga, S. C. (2017). A survey of intrusion detection in Internet of Things. *Journal of Network and Computer Applications*, *84*, 25–37. DOI: 10.1016/j.jnca.2017.02.009

Chapter 8
Smart Performance Analysis of Security-Enabled Mobile Apps for Banking Transactions

Navdeep Singh Sodhi
Chandigarh University, India

Sofia Singla
https://orcid.org/0009-0007-2123-4913
Chandigarh University, India

Isha Batra
Lovely Professional University, India

Arun Malik
Lovely Professional University, India

ABSTRACT

Banking sector has emerged into modern financial services and managing their transactions anywhere and anytime. This study explores the proposed frameworks based on the security of mobile banking systems focusing on multiple threats such as transaction fraud, phishing attacks, and unauthorized access. It focuses on the issues related to the user behavior that further leads into the vulnerabilities. Despite of focusing on the security level of mobile banking system, it is also important to focus on other measures like execution time consumption, storage, usage complexity and scalability. This study highlights the assessment of positive user experience towards the utilization of mobile banking applications. It is found that security enabled

DOI: 10.4018/979-8-3693-6925-8.ch008

framework can provide best and effective results to the users if all other parameters are considered by researchers. Overall, this study findings are able to promote a reliable and authentic banking application in an emerging financial economy.

INTRODUCTION

Trend of digital banking is growing rapidly all across the world. In the era of smartphones and digital services, banks have given their services as well as facilities with the help of authentic applications. It has made the lives of entire population easier and more reliable. Now-a-days, the need of visiting the banks has decreased and they can migrate the accounts and transfer their money with the means of these digital applications (Daniel et al., 2023). There is no need to step out towards the banks unnecessary. Further, digital transactions allow the users to access their banks at anytime from anywhere without any failure (Setiyawati & Bangkalang, 2022). Banks have ensured the multilevel security in order to protect the users as well as other stakeholders so that they can't be misguided by anyone. This impactful trend of digital transactions is highly followed by the users and their experience deals with the multiple factors like system design related factors, and user related factors. System related factors include functionality, usability, purpose and complexity whereas as user related factors deals with the moods, motivations, needs, expectations, and users' tendencies as shown in Figure.1. User experience basically includes the overall acceptance of the service and product provided by the banks. It ensures the customer satisfaction by further proceeding with the continuous intention and customer loyalty.

Figure 1. User experience for mobile applications

Mobile banking systems enable the users to perform non-financial as well as financial banking transactions from his or her current using corresponding devices. In the current era, users are experiencing multiple mobile applications like PhonePe, bank specific applications (Mobile Money, iMobile, AxisPay, SBIPay etc., GooglePay, Bharat Interface for Money (BHIM), UPI based mobile applications, Unsturctured Supplementary Service data, Immediate Payment Service, Short Messaging Service etc. for operations in credit card payments, loan payments, metro card recharge, mobile shopping etc (Lakshmi et al., 2019). In order to address the various issues along with the security of mobile banking applications, previous signature of attacks breaching the security of systems must be analyzed by the researchers. Banking systems including some particular functionalities are changing over the passage of time. New functionalities and appearances call new vulnerabilities as well (Wodo et al., 2021). Therefore, new vulnerabilities can be mitigated or discarded by having improvements in legislature means, education means, and technical means. Further, integration of finance and technology is to be considered as one of the great challenges by the researchers. Authentication factors ensure greater level of security as compared to other traditional static passwords (Khan et al., 2023). Financial transactions are increasingly facing the high cybercrimes. There is a great need to integrate the technologies for authentication of users with the mobile applications. It provides the best security to all legitimate users. Mobile banking applications aim on the information systems that includes the databases, network, stakeholders and stakeholders (A. Ojeniyi et al., 2019). Each and every business deal with the banking system so it is not wrong to say that if any of the threat involved in the banking system then it is directly lead to loss of that particular business. Security measures must be applied on the mobile applications to safeguard all kinds of businesses (Dzidzah et al., 2020) (Hanif & Lallie, 2021).

Problem Statement

This entire study is based on a problem faced by users while experiencing the services of mobile baking applications along with the security of their transactions. This study has highlighted the concerns regarding the mobile banking services including scalability, system complexity, storage requirements, and time consumption while experiencing the number of transactions. **"Mobile banking applications have faced the challenges like complex backend architectures, increased storage demands, long transaction processing times, that can further hinder the actual performance of applications, particularly during the peak time usage"**. Rather than just having advancements in the security of applications, there must be reliable intersection of underlined issues like scalability, storage optimization, usage complexity and execution time.

Objectives of book chapter:

The main objectives of the book chapter are:

- To review the existing research studies to analyze the security enabled frameworks based on mobile banking application.
- To discuss the challenges faced by researchers in order to propose frameworks for secure mobile banking applications.
- To compare the reviewed studies to evaluate their performance based on four different parameters like execution time, storage consumption, usage complexity and scalability.
- To discuss the results and find a research gap based on performance evaluated.

LITERATURE REVIEW

Many mobile applications or web applications are utilized for online ticket reservation, online gaming, and online banking systems by users all across the world. Researchers had discussed a monitoring system for increased detection of cyber-attacks on large scale systems. It had integrated the both unknown as well as known observers to detect numerous attacks like false data injection attack, replay attack, and covert attacks (Gallo et al., 2020). It is found that this particular research was totally based on the uncertainties. However, this particular research was totally based on the considered network and modules rather than considering any of the other dynamic metrics. Researchers have also discussed a framework based on the Near Field Communication for mobile payments. It has involved three different levels such as communication level, mobile application, and defense at hardware. This proposed framework has ensured the security using OCSP i.e. Online Certificate Status Protocol (Ahamad, 2021). It has considered cyber-attacks as well for the security of the mobile application and development of positive user experience. Further, a novel framework had proposed by researcher based on online banking systems utilizing Angular JS framework (Ghelani et al., 2022). It had discussed a minimal security based on biometric authentication. Further, other issues like grammatical, typographical and plagiarism problems were discussed in this particular framework (Ghelani et al., 2022). However, this research is not seen to be found very effective as it had not discussed anything related to smart security. Every day, new vulnerabilities are kept on entering into the market and breaching the security of applications. Therefore, no such feature related to smart security found in the framework. Due to increased vulnerabilities, cyber-attacks are kept on breaching the security of systems or applications. DDOS (Distributed Denial of Service) attack on the financial systems has become very common. An existing framework aimed at the detection of DDOS attack utilizing banking datasets. It had able to achieve 99.5% of the accuracy while uncovering the actual attempts of DDOS attack performed by intruders on financial systems (Islam et al., 2022). In the similar manner, many other cybercriminals are performing several potential attacks by taking advantage of today's technology (Almrezeq et al., 2023). This online dependency has created a great issue across the world. These kind of attacks are commonly introduced in smart (Isha et al., 2016). Moreover, cyber-attacks can be diminished as well as handled by utilizing Internet of Things. The entire ecosystem of smart cities embedded with sensors and other physical devices requires enhanced privacy and security (Khan et al., 2022). This can be resolved by using effective technological advancements in the mobile applications.

Among the payment wearable, security breaches have found as a biggest issue. In this context only, researchers had proposed a 3-authentication factor framework to mitigate the cyber threat as well as attacks. It had included AES (Advanced Encryption Standard) algorithm, ECDSA (Elliptic Curve Digital Signature Algorithm), and ECIES (Elliptic Curve Integrated Encryption Scheme) for highest and enhanced security in the framework (Bojjagani et al., 2023). It is found that this framework was able to treat the application with the maximum threats or possible attacks without any failure. It ensures more reliable performance while discussing the attacks on the mobile payments.

Researchers had discussed a secure framework utilizing YSGA (Yellow Saddle Goatfish Algorithm) in order to optimize frequent e-banking requests (Al-Zubaidie & Jebbar, 2024). This algorithm includes two phases i.e. 2PC (two-phase commit) and separates each type of the e-transaction into an isolated segment (Al-Zubaidie & Jebbar, 2024). It ensures the security in the framework with the help of applying authentication on each level. At the end, it is found that this particular algorithm is fast, secure and authentic for whole banking ecosystem. Online banking systems have faced an exponential advancement among the users in order to make any kind of transactions within a very small duration. In this aspect only, researchers have proposed a blockchain technology based framework for increased security as well as effective evaluation of performance metrics (Tsai et al., 2024). It has ensured a major positive impact on the security, transparency, immutability and efficiency of the online transactions and other financial services. In the terms of performance, cost effectiveness, security features, execution efficiency and extra certified devices are considered (Tsai et al., 2024). Further, integrating cloud based techniques in mobile applications introduce trustworthiness, consistent estimation independent rationality, and computational efficiency in the system. Research has discussed a new model based on the analysis of accuracy of data as well as experience of user (Shah et al., 2024). It is also found that it could also help in completion of concurrent tasks more effectively.

AI-based strategies have shaped the entire globe including healthcare and medicine, agriculture and food security, environmental conservation and climate change, economic development and finance development, workforce development and education, cyber security and banking systems. Reviewing about healthcare sector, wireless sensor network requires several secure strategies including authentication and authorization (Sama et al., 2022). Therefore, Body Sensor Network design can be utilized among mobile applications for consistent and trustworthy security and reliability for data communication. Researchers had proposed a novel AI based framework for banking systems (Pratap Singh Chauhan et al., 2024). It has mainly discussed about the mitigation of threats in order to address vulnerabilities growing across the globe. It is found that dynamic network mechanism has utilized in this framework. Performance has evaluated in terms of identification of threats, specified performance of threat, and scalability (Pratap Singh Chauhan et al., 2024). Researchers had found an effective result in terms of scalability as well. It is not wrong to state that mobile banking really need to consider a great attention for security of users. A user identification system must be developed to prevent or detect any kind of malware. MoBShield framework is proposed by researchers to prevent as well as detect the malware by using datasets generated from the numerous banking applications (Seraj et al., 2024). It had utilized XML (eXplainable Machine Learning) model to gather enhanced information from the systems related its security. It is found that this particular research is able to found highest accuracy as 99.7% by utilizing different approaches such as XML (eXplainable Machine Learning), CNN (Convolutional Neural Networks) and ANN (Artificial Neural Networks). Generation Z users are more effectively attempting the mobile payments for multiple reasons. Earlier, it was found that researchers were able to achieve at least 99.7% of the accuracy while having sufficient detection as well as prevention of malwares across the financial systems. However, suspicious detection ratio has also reached at its peak. Framework with the integration of blockchain and artificial intelligence has proposed (Alenizi et al., 2024). It is found that blockchain is able to store the data for each and every user whereas artificial intelligence is to make accurate decisions for improved security. This study discussed a fusion between the data security and cloud platforms along with nearly 98% of the accuracy (Alenizi et al., 2024). In the similar context, Artificial Intelligence can account for cybersecurity compliance, data protection, privacy legislation (Khan et al., 2025). Further, Artificial Intelligence for security can be considered as an effective technological advancement to protect the systems or mobile applications from any kind of cyber threats.

METHODOLOGY

In this study, a survey method has utilized to get the discussion and generate the accurate results regarding the usage of mobile application in the banking purposes by users as shown in Figure 2. This entire study is based on the secondary data gathered from the previous researchers. It has selected and reviewed all of the existing methodology as well as research studies based on the performance of the mobile banking applications. Further, it is seen that data gathered from the previous stage of selection and reviewing has shown multiple challenges among the studies. All of the challenges are analyzed and discussed so that users can be able to deal with these applications positively without any of the failure. Performance evaluation based on four parameters i.e. execution time, storage, complexity and scalability is seen for the entire study. This stage of the utilized methodology is considered as one of the most powerful stages among all as it deals with actual performance of mobile banking applications. These parameters are selected just to provide complete details to users regarding the actual storage consumption, scalable measures and many more. Further, this performance evaluation proceeded towards the results and discussion. Results and discussion has given a concise research gap regarding the positive usage of the mobile application. It won't be wrong to say that this stage is highly in assessing the mobile application in a meaningful manner. At the end, conclusion has generated for the entire study.

Figure 2. Methodology to assess the performance while utilizing mobile applications

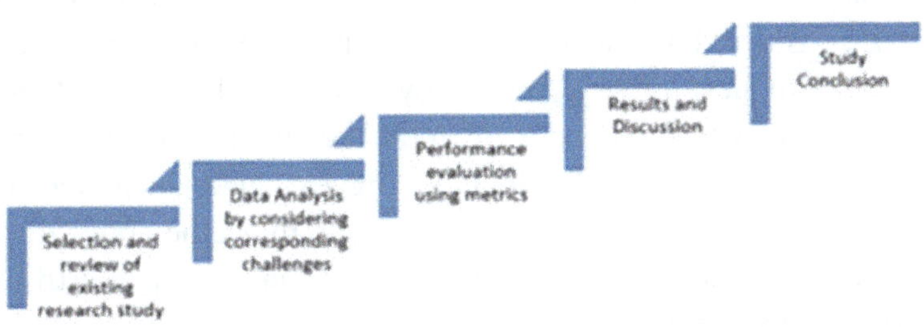

Challenges

In this section, multiple challenges related to existing frameworks have analyzed. Cyber attacks are considered as biggest challenge to handle while interacting with the online systems. In this similar manner, online financial systems are more frequently prone to the cyber-attacks and breach of users' sensitive data. Real time datasets are really required to analyze the actual fraud attempted on the application. However, it is very difficult to gather and track when researchers are working on the real-time datasets. Mobile payments have come to existence very frequently. Users are becoming more habitual to perform mobile payments rather than cash payments even for minimal bills. That's why, secure mobile payments have become a biggest challenge across the entire banking sector. In this context, blockchain technology can be one of the most beneficial as it can introduce dynamic networks (Bojjagani et al., 2023). Smart contract framework is highly utilized at all levels across the entire globe. It is basically directly executed by the program or systems rather than showing any of manual entry or protocol of performing transactions. Further, microsegmentation is a biggest challenge faced by researchers in the overall proposal of framework. It has discussed the challenges in terms of execution time, speed, complexity, security, storage and many more (Al-Zubaidie & Jebbar, 2024). However, existing framework is able to ensure the security of the users by providing reliable execution of all terms and conditions, transparent as well as immutable records of transaction, and cryptographic security including end users' authentication on application while going through e-transactions (Al-Zubaidie & Jebbar, 2024).

Despite of discussing all secure an efficient online financial systems or applications, there are still number of challenges reside that need to be consider further in the future researchers. Scalability is a biggest challenge that must be involved in the proposed frameworks. Frameworks are seen to be incapable while meeting with the increased traffic flow of users. Application starts on demanding high processing period (Tsai et al., 2024). Moreover, blockchain is applied on each framework but discussing about the security of particular online financial activity involved in the system must be considered. It is seen that some of the common attacks like credential sniffing, cross-site request forgery, firewall bypass, eavesdropping, DNS spoofing, Ransomware, Zero-Day Exploit and many more are unable to detect by researchers (Pratap Singh Chauhan et al., 2024). So, it is a biggest challenge to detect each type of cyber-attack on any kind of online financial system. In the above discussion of existing frameworks, it is seen that integration of blockchain and artificial intelligence can provide better result along with 98% of accuracy (Alenizi et al., 2024). But still, scalability is again considered as one of the biggest challenges to consider in the smart and secure financial sector.

Performance Evaluation of Existing Frameworks

This section shows the parameters like execution time, storage, complexity and scalability to evaluate the performance of existing frameworks.

Table 1. Performance evaluation based on multiple parameters

Ref	Year	Execution Time	Storage	Complexity	Scalability
Gallo et al., 2020	2020				
(Ahamad, 2022)	2022	✓	✓	✓	
Ghelani et al., 2022	2022		✓		
Islam et al., 2022	2022	✓		✓	
Bojjagani et al., 2023	2023	✓	✓	✓	
Al-Zubaidie & Jebbar, 2024	2024	✓	✓	✓	
Tsai et al., 2024	2024	✓	✓	✓	
Pratap Singh Chauhan et al., 2024	2024				✓
Seraj et al., 2024	2024	✓			
Alenizi et al., 2024	2024	✓	✓	✓	

RESULTS AND DISCUSSION

In this study, results are found based on four different parameters such as execution time, storage, complexity and scalability as shown in Figure 3. The above reviewed literature has shown that researchers are showing great interest in providing the enhanced security to all of the users without any failure. No doubt, security is highly demandable to develop maximum user experience towards the mobile banking applications but still other parameters need to consider as well. It is found that execution time of proposed framework is considered by the users more effectively. Among all of the reviewed studies, it is seen that seventy percentile of the researchers are considering the actual consumption of the time. Further, storage and usage complexity are seen to be considered in the sixty percent of the cases. However, considering about the scalability, it is seen that very few researchers had discussed. Discussing about these results, it is not wrong to say that these parameters must be evaluated in all of the cases even if researchers are able to secure enhanced security in the system. It is found that all of the parameters i.e. execution time, storage, usage complexity and scalability must be considered as along with the security of the mobile application. If these parameters along with the security of users are able to provide best results, then users can be able to achieve best and enhanced user experience from the mobile banking applications.

Figure 3. Performance evaluation of existing frameworks

CONCLUSION

Reviewed studies found that digital mobile banking applications need to improve user related services enabled security and other metrics. It is concluded that there is a high requirement of integration between secure digital finance and application related stability for increased user experience. This study aims to assess the overall performance of mobile banking application along with the security integrated among them. In particular, results have found a clear research gap of not considering other parameters like execution time consumption, storage, usage complexity and scalability along with the security measures. Due to this major issue, mobile banking applications are not still considered as best platform for sensitive transactions globally. Furthermore, enhanced technologies like block chain, artificial intelligence, Internet of Things and many more can be utilized for smart financial ecosystem in the future.

REFERENCES

Ahamad, S. S. (2021). A novel NFC-based secure protocol for merchant transactions. *IEEE Access : Practical Innovations, Open Solutions, 10*, 1905–1920. DOI: 10.1109/ACCESS.2021.3139065

Al-Zubaidie, M., & Jebbar, W. (2024). Transaction security and management of blockchain-based smart contracts in e-banking-employing microsegmentation and yellow saddle Goatfish. *Mesopotamian Journal of CyberSecurity, 4*(2), 1–19. DOI: 10.58496/MJCS/2024/005

Alenizi, A., Mishra, S., & Baihan, A. (2024). Enhancing secure financial transactions through the synergy of blockchain and artificial intelligence. *Ain Shams Engineering Journal, 15*(6), 102733. DOI: 10.1016/j.asej.2024.102733

Almrezeq, N., Humayun, M., Alruwaili, M., Alanazi, S., & Jhanjhi, N. Z. (2023). Cyber security attacks and challenges in Saudi Arabia during COVID-19. *International Journal of Computer Science & Network Security, 23*(10), 179–187.

Bojjagani, S., Seelam, N. R., Sharma, N. K., Uyyala, R., Akuri, S. R. C. M., & Maurya, A. K. (2023). The use of iot-based wearable devices to ensure secure lightweight payments in fintech applications. *Journal of King Saud University. Computer and Information Sciences, 35*(9), 101785. DOI: 10.1016/j.jksuci.2023.101785

Chauhan, R. P. S., Sonker, S. K., Kaur, M., Sharma, C., Singh, R., & Singh, R. (2024, March). Optimizing IoT Threat Mitigation with Artificial Intelligence in Banking: A Multi-Objective Approach. In *2024 2nd International Conference on Disruptive Technologies (ICDT)* (pp. 296-301). IEEE.

Dzidzah, E., Owusu Kwateng, K., & Asante, B. K. (2020). Security behaviour of mobile financial service users. *Information and Computer Security, 28*(5), 719–741. DOI: 10.1108/ICS-02-2020-0021

Gallo, A. J., Turan, M. S., Boem, F., Parisini, T., & Ferrari-Trecate, G. (2020). A distributed cyber-attack detection scheme with application to DC microgrids. *IEEE Transactions on Automatic Control, 65*(9), 3800–3815. DOI: 10.1109/TAC.2020.2982577

Ghelani, D., Hua, T. K., & Koduru, S. K. R. (2022). A Model-Driven Approach for Online Banking Application Using AngularJS Framework. *American Journal of Information Science and Technology, 6*(3), 52–63.

Hanif, Y., & Lallie, H. S. (2021). Security factors on the intention to use mobile banking applications in the UK older generation (55+). A mixed-method study using modified UTAUT and MTAM-with perceived cyber security, risk, and trust. *Technology in Society*, *67*, 101693. DOI: 10.1016/j.techsoc.2021.101693

Isha, L., A. K., & Kumar, S. (2016, August). Layer based security in Internet of Things: current mechanisms, prospective attacks, and future orientation. In International Conference on Smart Trends for Information Technology and Computer Communications (pp. 896-903). Singapore: Springer Nature Singapore.

Islam, U., Muhammad, A., Mansoor, R., Hossain, M. S., Ahmad, I., Eldin, E. T., Khan, J. A., Rehman, A. U., & Shafiq, M. (2022). Detection of distributed denial of service (DDoS) attacks in IOT based monitoring system of banking sector using machine learning models. *Sustainability (Basel)*, *14*(14), 8374. DOI: 10.3390/su14148374

Khan, A., Jhanjhi, N., Abdulhabeb, G. A. A., Ray, S. K., Ghazanfar, M. A., & Humayun, M. (2025). Securing IoT Devices Using Generative AI Techniques. In *Reshaping CyberSecurity With Generative AI Techniques* (pp. 219–264). IGI Global.

Khan, A., Jhanjhi, N. Z., & Humayun, M. (2022). The role of cybersecurity in smart cities. In *Cyber Security Applications for Industry 4.0* (pp. 195–208). Chapman and Hall/CRC. DOI: 10.1201/9781003203087-9

Khan, H. U., Sohail, M., Nazir, S., Hussain, T., Shah, B., & Ali, F. (2023). Role of authentication factors in Fin-tech mobile transaction security. *Journal of Big Data*, *10*(1), 138. DOI: 10.1186/s40537-023-00807-3

Lakshmi, K. K., Gupta, H., & Ranjan, J. (2019). UPI based Mobile Banking Applications – Security Analysis and enhancements. *2019 Amity International Conference on Artificial Intelligence (AICAI)*, 1–6. DOI: 10.1109/AICAI.2019.8701396

Ojeniyi, J. A., Edward, E. O., & Abdulhamid, S. M. (2019). Security risk analysis in online banking transactions: Using diamond bank as a case study. *International Journal of Education and Management Engineering*, *9*(2), 1–14. DOI: 10.5815/ijeme.2019.02.01

Sama, N. U., Zen, K., Humayun, M., Jhanjhi, N. Z., & Rahman, A. U. (2022). Security in wireless body sensor network: A multivocal literature study. *Applied System Innovation*, *5*(4), 79. DOI: 10.3390/asi5040079

Seraj, S., Sadiq, A. S., Kaiwartya, O., Aljaidi, M., Konios, A., Ali, M., & Abazeed, M. (2024). MoBShield: A Novel XML Approach for SecuringMobile Banking. *Computers, Materials & Continua*, *79*(2), 2123–2149. DOI: 10.32604/cmc.2024.048914

Setiyawati, N., & Bangkalang, D. H. (2022, September). The comparison of evaluation on user experience and usability of mobile banking applications using user experience questionnaire and system usability scale. In *Proceedings* (Vol. 82, No. 1, p. 87). MDPI. DOI: 10.3390/proceedings2022082087

Shah, S. A. A., Ullah, A., Subhan, F., Jhanjhi, N. Z., Masud, M., & Alqhatani, A. (2024). Truth discovery for mobile workers in edge-assisted mobile crowdsensing. *ICT Express*, *10*(5), 1087–1093. DOI: 10.1016/j.icte.2024.06.007

Tsai, C. H., Liou, D. K., & Lee, H. L. (2024). Blockchain-supported online banking scheme. *Egyptian Informatics Journal*, *27*, 100516. DOI: 10.1016/j.eij.2024.100516

Wodo, W., Stygar, D., & Blaskiewicz, P. (2021, July). Security Issues of Electronic and Mobile Banking. In *SECRYPT* (pp. 631-638).

Chapter 9
Innovative Methods for Mobile App Evaluation Using Smart Assessment Tools

Jimmy Singla
https://orcid.org/0000-0003-2077-1622
Lovely Professional University, India

ABSTRACT

In this chapter, the methods to mobile app evaluation through smart assessment tools are explored. Traditional methods are not enough to capture user experience and app performance. A collection of smart methodologies that support data analytics and machine learning to enhance the evaluation process are presented here. The smart assessment tools make use of combination of real-time data and user centered design principles. With this combination, these smart assessment tools provide deeper perceptions into app usability and functionality. These methods further result in increased user satisfaction. There are many case studies where these methods are applied across many industries such as healthcare app evaluation, E-learning platforms and E-Commerce mobile apps.

1. INTRODUCTION

Now a days, mobile apps play a very important role in our lives. These apps make our lives easy and convenient. Many tasks can be completed very easily with the help of mobile apps. These tasks include ordering food, managing finances, tracking fitness, connecting with friends etc. The number of apps is available in millions

DOI: 10.4018/979-8-3693-6925-8.ch009

now and this number is increasing day by day. So, there is high competition in the market and user's expectation is also increasing. This is the reason that it becomes mandatory to evaluate the mobile apps and thus to know the user's interactions with apps. A sample of mobile apps is shown in figure 1.

Figure 1. Mobile Apps

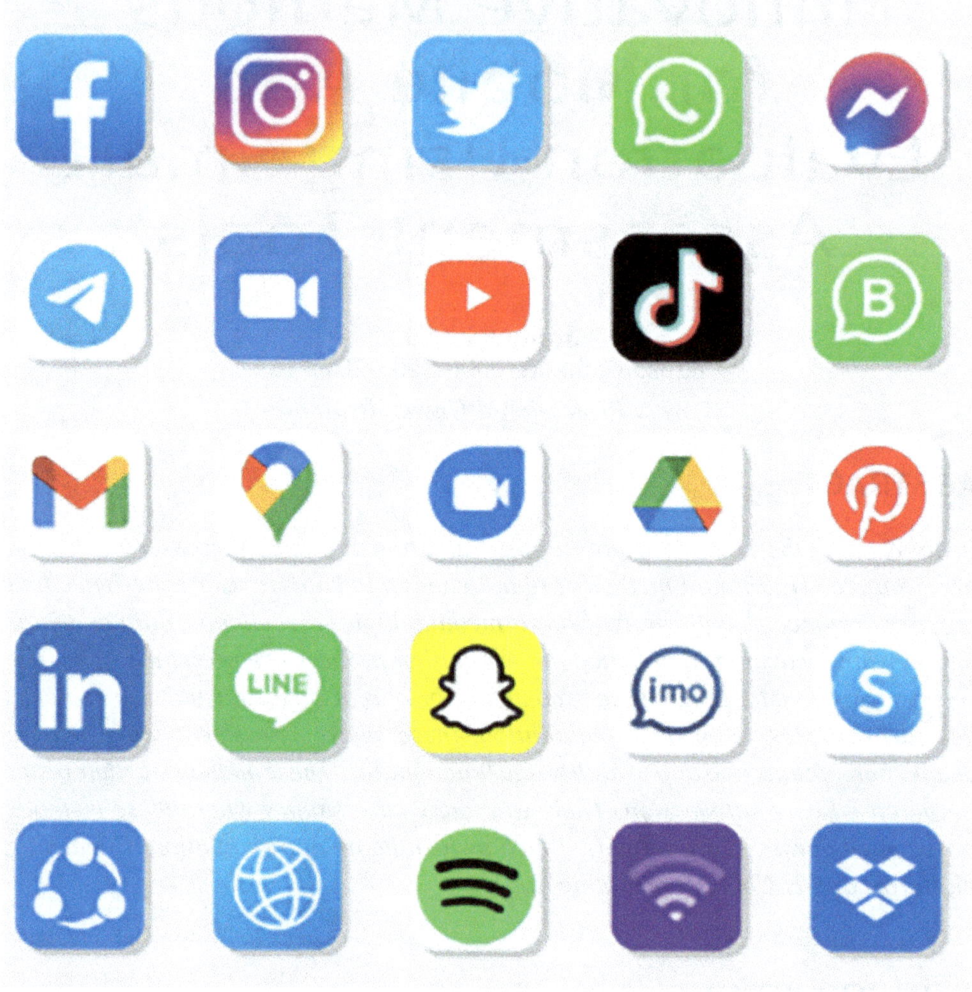

Traditional methods evaluate mobile apps in terms of number of downloads, ratings and reviews. With this, one can come to know about how an app is doing but it does not give the complete picture. The understanding is required on what users actually experience while using the apps. The developers must know the user's experiences with the mobile apps. This results in the need of smart evaluation methods for mobile apps. These smart evaluation methods are based on advanced technologies such as data analytics and user feedback systems. These methods provide real time perceptions into how an app is doing. Otherwise, in traditional methods, it depends on user's reviews and ratings. So, one should wait for user's reviews and ratings in traditional methods. On the other hand, in smart methods, it considers interaction of people with apps right away. The advanced technologies for smart evaluation methods for mobile apps are shown in figure 2.

Figure 2. Advanced technologies for smart evaluation methods

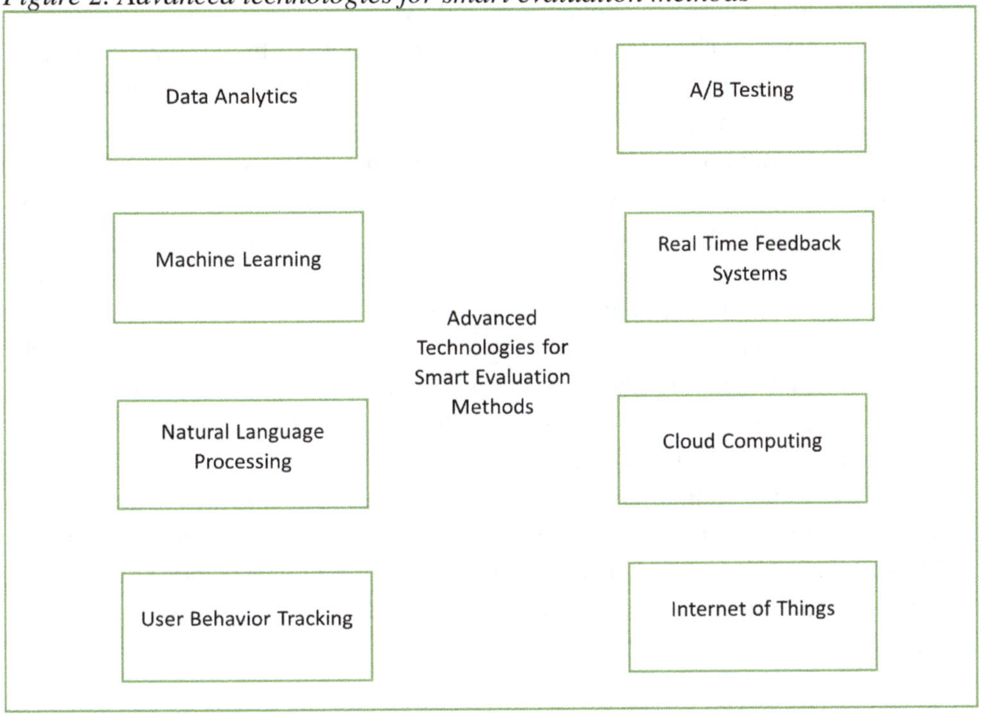

Data analytics techniques are used to find patterns from large volumes of user's data. Machine learning algorithms enable systems to learn from data and improve the performance with the passage of time. Natural language processing helps to analyze and understand user feedback through text. User tracking behavior tools help to monitor the interactions of users with apps. A/B testing helps to test different versions of mobile apps. With this, one can know which version is better based on the response of the user. Real time feedback systems give immediate feedback within the app and quick adjustments are made based on user's experiences. Cloud computing supports scalable data storage and processing of units. Internet of things provide integration with connected devices for better user experiences. The above-mentioned technologies are combined together to produce an effective framework for mobile app evaluation.

2. IMPORTANCE OF MOBILE APP EVALUATION

Mobile app evaluation is important because of several factors such as user experience, performance optimization, security assessment, competitive advantage, feedback for improvements, monetization strategies, compliance & standards and market adaptation. These are explained as follow:

User experience – Usability issues can be identified with app assessment and evaluation. It ensures smooth user experience which will result in higher user retention.

Performance optimization – Performance bottlenecks can be identified with app assessment and evaluation. It can identify crashes and slow load times.

Security Assessment – The vulnerabilities can be identified with regular assessment and evaluations. It ensures data protection and user trust.

Competitive advantage – Apps pros and cons with respect to the competitors can be identified and further enhancements can be done.

Feedback for improvements – Evaluations and assessments provide user feedback. It focuses on what user actually wants. It guides on future updates as well.

Monetization strategies – it ensures the integration of in-app purchases and ads with any compromise with user's experience.

Compliance and Standards – Evaluation and assessment ensures that the app complies with regulations and standards of industry and further reduces legal risks.

Market adaptation – Market trends and user needs always keep on changing. Continuous evaluation and assessment help apps adapt to these changing circumstances effectively.

3. OVERVIEW OF SMART ASSESSMENT TOOLS

Smart assessment tools provide real time feedback and they also help to streamline the flow of assessment for mobile apps. The features, benefits and examples of smart assessment tools are mentioned in Table I.

Table 1. Features, benefits and examples of Smart Assessment Tools

Smart Assessment Tools		
Features	**Benefits**	**Examples**
Automated Grading	Efficiency	TestFlight
Adaptive Learning	Immediate Feedback	Firebase Test Lab
Data Analytics	Personalization	Appium
Multimedia Integration	Enhanced Engagement	UserTesting
Collaboration Features	Scalability	Lookback

The features of Smart assessment tools include automated grading, adaptive learning, data analytics, multimedia integration and collaboration features. Automated grading saves time with the help of algorithms. With this, human error is minimized. It also ensures fair assessments. With adaptive learning, difficulty can be adjusted based on user's response. With data analytics, performances can be monitored. Reports on strengths and weaknesses can be generated with data analytics. Multimedia integration makes assessments more interactive and engaging. It supports video, audio and many interactive scenarios. Collaboration features work on critical thinking skills. It supports various collaborative tools which helps in assessments of mobile apps.

The benefits of smart assessment tools are efficiency, immediate feedback, personalization, enhanced engagement and scalability. It saves time with automation of grading and analysis part. With smart assessment tools, immediate feedback is always received. Personalization involves adapting contents based on performance and interaction. Enhanced engagement refers to interactive features such as assessment in the form of games and quizzes. It also makes assessment enjoyable.

The examples of Smart Assessment tools are TestFlight, Firebase Test Lab, Appium, UserTesting and Lookback. TestFlight is Apple's official tool for beta testing iOS apps. In this tool, users are invited by developers for testing their apps. Users feedback is collected and issues are tracked before the app is finally released. Its working is shown in figure 3. Firebase Test Lab is a cloud-based service offered by Google that allows developers to test their mobile apps on a wide range of devices and configurations. It helps identify issues and improve app quality before release. Appium is an open-source automation tool designed for testing mobile applications. It allows developers to write tests for both native and hybrid mobile apps across various platforms, including iOS and Android. UserTesting is a user experience (UX) stage that enables organizations to gather feedback from real users interrelating with their products, including mobile apps. It helps teams in identifying user behavior and improve overall usability.

Figure 3. Working of TestFlight

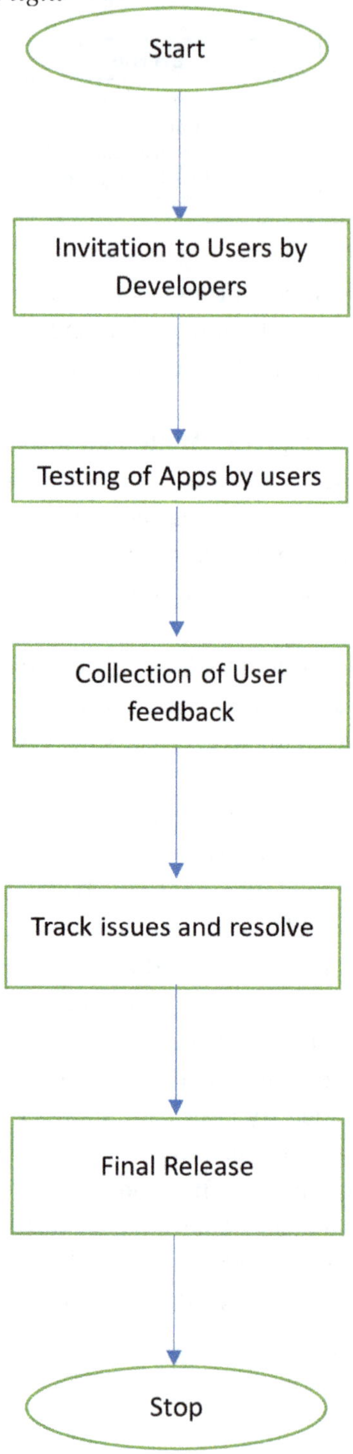

4. TRADITIONAL EVALUATION METHODS

The traditional evaluation methods for mobile apps user testing, surveys and questionnaires, heuristic evaluation and performance testing. It is shown in figure 4.

Figure 4. Traditional Evaluation Methods

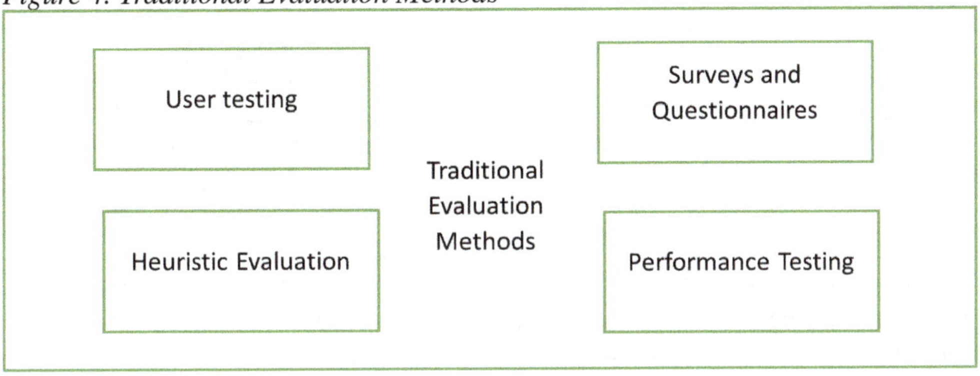

User testing involves observing real users who are interacting with the apps, identifying different issues with usability and finding weak areas which can be improved. Surveys and questionnaires include gathering subjective feedbacks and opinions from different users about their satisfaction. In heuristic evaluation, certain established usability principles are used to identify the usability issues. In performance testing, the working of apps is checked under various different circumstances or contexts and it ensures that the app performs well in these contexts.

5. LIMITATIONS OF CONVENTIONAL METHODS

Each conventional method has some limitations. These limitations are shown in Table II.

Table 2. Limitations of Conventional Methods

S No.	Conventional Method	Limitations
1.	User Testing	Limited Sample Size Observer effect Resource-intensive
2.	Surveys and Questionnaires	Response Bias Limited Depth Low Response Rate
3.	Heuristic Evaluation	Subjectivity Expertise Required Potential for overlooked issues
4.	Performance Testing	Ideal Conditions Technical Focus High Complexity

The conventional methods discussed in this chapter are user testing, surveys and questionnaires, heuristic evaluation and performance testing. These conventional methods have their limitations. The user training method has limitations such as limited sample size, observer effect and its resource intensive nature. In user testing, a small group of participants are always involved. So, it can not be able to represent broad user base. In this user testing, there is one major problem that the user can change his behavior when he knows he is being watched. It effects the authenticity of user's interactions with apps. One another limitation of user testing is, it required significant time and effort for session assistance and analysis. Another method discussed here in this chapter is surveys and questionnaires, its limitations are response bias, limited depth and low response rate. Sometimes users are not giving honest results and responses. They are giving responses which social media desires. Another limitation of this method is its limited depth. It can't find the reasons behind the user's behavior and interactions. Many people also don't bother to fill the surveys and questionnaires. So, sufficient numbers of responses may not be received. It always suffers from incompleteness in data. Heuristic evaluations suffer from subjectivity, requirement of expertise and potential for overlooked issues. In this method, results can vary because all the experts may have different expertise. All experts must have knowledge of usability principles which is not always fulfilled. It also misses contextual issues. The next method which is performance testing also have some limitations. These are ideal conditions, technical focus and high complexity. The performance testing is not performed in real world contexts. It is always conducted in controlled environment. This is the reason that it leads to misleading results. This method can also miss the usability issues effecting the real-world usage because it gives priority to technical performance as compared to usage performance. This method also requires advanced technical knowledge and tools. It is not always possible to provide access of these tools to all the teams.

6. INNOVATIVE METHODS IN MOBILE APP EVALUATION

The innovative methods in mobile app evaluation are remote usability testing, eye-tracking technology, affective computing, in-app feedback tools, gamification in testing, machine learning for predictive analysis and augmented reality testing. Each method with its benefits is shown in Table III.

Table 3. Innovative Methods in Mobile App Evaluation

S No.	Method	Benefits
1	Remote usability testing	• Cost effective • Access to diverse participants • Flexibility in scheduling • Quick turnaround • Immediate insights • Rich data collection
2	Eye-tracking technology	• Precise attention mapping • Identification of usability issues • Quantitative and Qualitative data • Real time insights • Innovative presentations of findings
3	Affective computing	• Enhanced user understanding • Improved user engagement • Real time feedback • Engaging user interfaces
4	In-app feedback tools	• Iterative development • Improved retention rates • Data-driven decisions • Contextual understanding • Enhanced user experiences
5	Gamification in testing	• Better data quality • Real-time feedback • Increased engagement
6	Machine learning for predictive analysis	• Enhanced accuracy • Automated insights • Scalability • Continuous learning
7	Augmented reality testing	• Immediate feedback • Creative problem solving • Increased user engagement • Rich data collection

Remote usability testing is one of the prominent methods to collect user feedback and make improvements in app design (Hammontree et al.,1994). The right way to conduct the remote usability is to first define objectives and then selection of right tool takes place. After this test scenarios are developed, participants are recruited. Then final test is conducted and results are analyzed. In the end, follow-up takes place. Eye-tracking technology involves special hardware and software for the purpose of monitoring where users watch on the screens (Punde et al., 2017). It focuses on how users navigate the app, which elements attract the users and where they find difficulties. Affective technologies include emotion recognition technologies to increase the performance of mobile app evaluation. In this technique, on the basis of analysis of user's facial expressions, developers can identify emotional states of users while they navigate the app (Lasen, 2004). In-app feedback tools are crucial for gathering user insights directly within mobile applications, enabling developers to understand user experiences and identify improvement areas in real time. Features like surveys, bug reporting, and NPS (Net Promoter Score) allow users to easily share their thoughts while engaged with the app, leading to more accurate and actionable data (Scherr et al., 2022). Gamification in testing includes game like test elements for better user engagement. For this purpose, features like rewards points and challenges are added for more interactive environment for users (Cacciotto et al., 2021). Machine learning for predictive analysis in mobile app evaluation leverages algorithms to analyze user behavior patterns and predict future actions, such as user retention or potential churn. By examining historical data, these models can identify trends and suggest enhancements to app features or user experiences (Pandey et al., 2020). Augmented reality (AR) testing for mobile apps includes assessing how well AR structures work in real-world environments, ensuring seamless integration of digital elements with physical surroundings. This testing assesses user interactions, spatial recognition, and the complete immersive experience. By recognizing usability problems and improving performance, developers can improve user visit and gratification with AR applications (Davidavičienė et al., 2021).

7. CONCLUSION

The innovative methods for mobile apps evaluation using smart assessment tools are discussed in this chapter. From this discussion, it is concluded that the use of mobile applications is increasing day by day and user satisfaction is very essential component of mobile apps assessment. Further the need for effective evaluation methods is very crucial. The methodologies discussed in this chapter, not only enhance the superiority and serviceability of mobile applications but also increase a more appealing and adapted user involvement. By concentrating on user-centric design and integrating diverse perceptions, developers can confirm that their applications meet the varied requirements of their audience, ultimately leading to greater user satisfaction and trustworthiness.

8. FUTURE DIRECTIONS

In future, the assessment and evaluation methods for mobile apps can be further improved by integration of artificial intelligence, real time feedback mechanism and user centric designs. The work can be done to enhance user experience, improve app performance and drive engagement. Moreover, combining augmented reality and virtual reality in future can enable better evaluations of user interactions within apps.

Users from different areas can also be considered so that different inspirations can also be identified on app usage and assessment. Blockchain technology can be implemented to ensure the authenticity of user feedbacks. It would increase trust in the process of mobile app evaluation.

9. REFERENCES

Cacciotto, F., Fulcini, T., Coppola, R., & Ardito, L. (2021, April). A metric framework for the gamification of web and mobile gui testing. In *2021 IEEE International Conference on Software Testing, Verification and Validation Workshops (ICSTW)* (pp. 126-129). IEEE. DOI: 10.1109/ICSTW52544.2021.00032

Davidavičienė, V., Raudeliūnienė, J., & Viršilaitė, R. (2021). Evaluation of user experience in augmented reality mobile applications. *Journal of Business Economics and Management*, *22*(2), 467–481. DOI: 10.3846/jbem.2020.13999

Hammontree, M., Weiler, P., & Nayak, N. (1994). Remote usability testing. *interactions, 1*(3), 21-25.

Lasén, A. (2004). *Affective technologies—emotions and mobile phones*. The Digital World Research Centre.

Pandey, M., Litoriya, R., & Pandey, P. (2020). Applicability of machine learning methods on mobile app effort estimation: Validation and performance evaluation. *International Journal of Software Engineering and Knowledge Engineering*, *30*(01), 23–41. DOI: 10.1142/S0218194020500023

Punde, P. A., Jadhav, M. E., & Manza, R. R. (2017, October). A study of eye tracking technology and its applications. In *2017 1st International Conference on Intelligent Systems and Information Management (ICISIM)* (pp. 86-90). IEEE. DOI: 10.1109/ICISIM.2017.8122153

Scherr, S. A., Ter-Tovmasyan, M., Neugebauer, F., Hupp, S., & Elberzhager, F. (2022). The way it made me feel-Creating and evaluating an in-app feedback tool for mobile apps. *J. Ubiquitous Syst. Pervasive Networks*, *17*(1), 27–34.

Zhang, C., Zhang, X., & Halstead-Nussloch, R. (2014). ASSESSMENT METRICS, CHALLENGES AND STRATEGIES FOR MOBILE HEALTH APPS. *Issues in Information Systems*, *15*(2).

Chapter 10
Quantum Machine Learning in Cloud-Based Security Services and Applications

Ravikumar Ch
https://orcid.org/0000-0003-0809-5545
Sreenidhi University, Hyderabad, India

K. Raghavendar
https://orcid.org/0000-0001-5554-2614
TKR College of Engineering, India

Isha Batra
Lovely Professional University, India

Ghadah Naif Alwakid
College of Computer and Information Sciences, Jouf University, Saudi Arabia

Khulud Salem S. Alshudukhi
College of Computer and Information Sciences, Jouf University, Saudi Arabia

ABSTRACT

The integration of Quantum Machine Learning (QML) into cloud-based security services represents a transformative shift in cybersecurity. By leveraging quantum computing's ability to process data exponentially faster than classical systems, QML enhances cloud security with real-time threat detection, anomaly recognition, and intrusion prevention. It introduces quantum-inspired algorithms, quantum versions of classical algorithms, and hybrid quantum-classical models, each offering unique

DOI: 10.4018/979-8-3693-6925-8.ch010

advantages in processing speed and scalability. However, the rise of quantum computing also threatens conventional encryption methods like RSA and ECC, necessitating the development of quantum-resistant encryption techniques. Despite challenges posed by current quantum devices, QML's potential to revolutionize cloud security is immense, providing faster, more efficient, and robust protection against evolving cyber threats.

1. INTRODUCTION

Quantum computing capabilities in cloud computing have preconditioned major breakthroughs in developing quantum computing which is a fast-evolving branch of computer science (Malhotra, 2022). The unification between cloud services and quantum computers has brought about new possibilities, especially in the field of quantum machine learning (QML) that has achieved outstanding developments over the past years. This paper investigates the symbiotic nature between cloud platforms and quantum computing, in particular, the way such compilation can bring quantum resources closer to the machine learning process. The concept of quantum computing has also been opened up because cloud-based quantum computing has removed common impediments, including the requirement of expert machinery and knowledge Whyte, 2024).

What is more, the growth of volume and complexity of data with an exponential scale has only boosted the development of machine learning, which still serves as the foundation of artificial intelligence (AI) (Anagolum et al., 2024). Machine learning Quantum computing in clouds in a way is transforming the form of data analytics and decision-making, which may even be subjective to quantum-improved algorithms over their classical analogs. The paradigm change between quantum computing and machine learning announces a new bountiful era of inventions in AI and data-driven technologies.

Figure 1. Quantum Computing and Cloud Computing

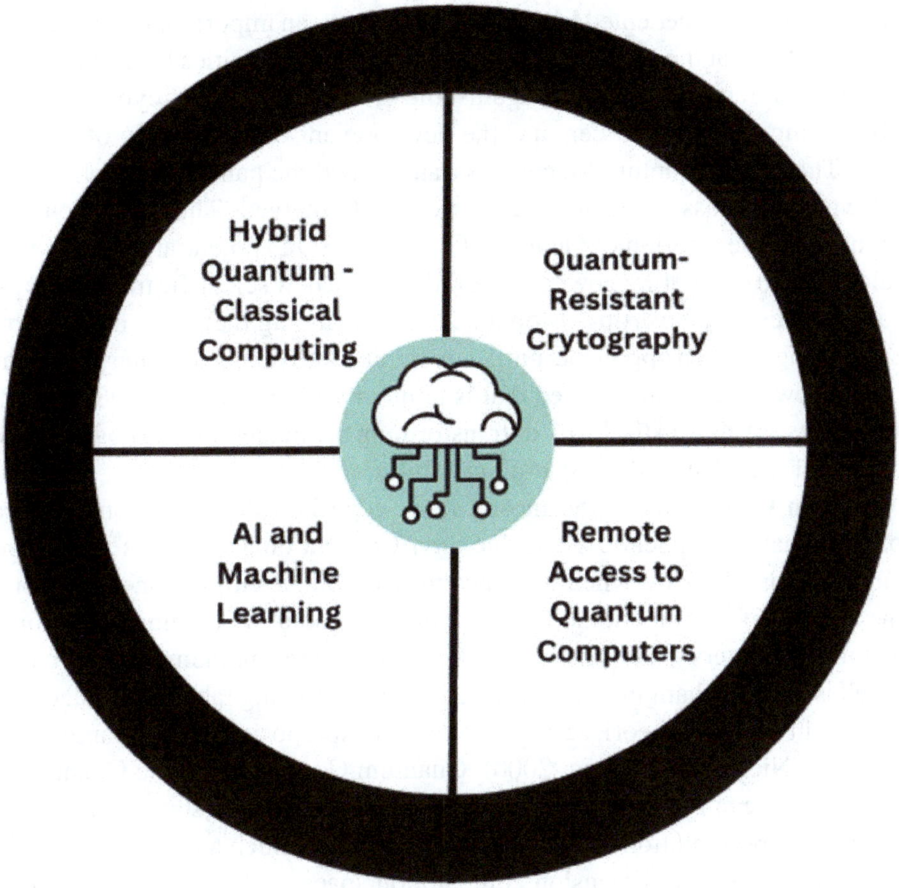

It can be concluded that the combination of quantum computing and machine learning is a technological wonder that will represent a revolution. Blending Quantum Machine Learning (QML) into cloud-based security services and apps is a playing field this chapter will explore, breaking down the opportunities, obstacles, and roads forward in this exciting innovation. Computing at its present position is facing a new paradigm shift. Computer technology, the next generation, along with the doubling of computational research, promises to usher in a golden age of discovery, problem solving and creativity (Feynman, 1981). This paper sets out on a journey of discovery into these quick transitions, exploring the nature in which the computational science is changing and how it has had far reaching implications on different areas of science.

The overwhelming force of networking has transformed the normal lives, as it has created perfect communication, cooperation and information access. Nonetheless, along with this unprecedented interconnection comes an important issue of cybersecurity. Granted that the internet facilitating all kinds of activities lacks the powerful security that it requires to protect against the growing menace of cyber-attacks. At the beginning of the 20 th century, the development of the Theory of Relativity and the Theory of Quantum Mechanics transformed the panorama of physics, and the dawn of the possible technological potential emerged. The revolutionary era brought about the creation of Quantum Computing (QC) potentially being the next development in computing power. Elaborating that a new scientific truth never gains the ascendancy by converting its opponents and bringing them into the light but it rises by going through opponents until they finally outlive their counterparts and die off and a new generation is raised that is familiar with it (Planck, 1998). Quantum technologies are also on the brink of transforming the modern technologies, leading to the era of the advanced computational abilities.

Quantum Computing is the most cutting-edge knowledge in computing and implies unmatched quickness and lucidity. Conventional practices of computer security are problematic when it comes to decryption with exponential increase in the amount of computational power. A quantum solution coming in the form of Quantum Computers which can solve exponentially large mathematical problems in parallel, is a game-changing solution. Quantum Computing takes advantages of the postulates in quantum theories and makes use of superposition and entanglement in computing (Nielsen & Chuang, 2000). Quantum Computers utilise Quantum Bits (qubits) which can be represented in a superposition of both states at once corresponding to representation of complex configurations. Such a characteristic allows qubits to develop multidimension computation spaces, which offers new insights into complex tasks.

The looming emergence of Quantum Computing is creating a batch of new forms of devices that are described as Noisy Intermediate Scale Quantum (NISQ) devices. Although still not being error-corrected or noise-free, these devices will spur the demand of having increased computing power (Preskill, 2018).

1.1 A Brief History of Quantum Computing: Milestones and Achievements

Quantum communication theory with its origins in the famous suggestion by Albert Einstein of the concept of the existence of a spooky action at a distance, has given rise to many new innovations in the area of quantum computing. The idea opened the door to the creation of quantum information theory, whose breakthroughs were characterized by milestones. In 1976 Roman Stanislaw Ingarden, and in 1981,

Richard Feynman headlined a keynote address by the name Simulating Physics with Computers in which he indicated the prospects of quantum computers to provide solutions that modern classical computers could not (Ash-Saki, Alam, & Ghosh, 2020). In 1992, another of the first practical quantum algorithms, the DeutschJosza algorithm, became the realization of a quantum algorithm that might run substantially faster than its classical equivalent, a milestone in the history of quantum computers.

In 1996 Lov Grover proposed quantum algorithm to search database which demonstrated further the potential of quantum computing as still being revolutionary. This development prompted a further investigation into quantum technologies and major milestones can be found in the years that followed. As an example, the Quantum Computation Roadmap created in 2002 also served to guide the researchers as well as defining the future direction of quantum computing. D-Wave Systems sold the first commercial quantum computer in 2011 and in 2012 1QB Information Technologies (1QBit) developed the software to interface with quantum computers, both noteworthy events. Recently, in 2017, Chinese physicists succeeded in the quantum communication of single-photon qubit via 1400 kilometers in the quantum teleportation, a significant performance in the history of quantum computing (Ayanzadeh, Nguyen, & Proctor, 2023).

Among the most outstanding developments was the quantum supremacy acquired by Google in the year 2019, processing a task 200 seconds that would require a supercomputer 10,000 years. This was a change of paradigm in computing power, showing that quantum computing has unlimited potential to transform computer science (Balewski, Mukherjee, Perez, & Haffner, 2024). The reaction of IBM to this breakthrough gave rise to set of recommendations that seek to improve the power of supercomputers, yet another point that underlines the rise in competition and the rising technology in quantum computing.

1.2 Quantum Machine Learning and Any Cloud-Based Security

Quantum machine learning (QML) is an incredible merger of quantum technologies and machine learning as the dawn of novel computational technology. Quantum bits (qubits) are capable of storing much more information than their classical counterparts, and superposition and quantum entanglement enable data processing on quantum computers exponentially faster than currently possible with traditional systems. Quantum computers are able to complete some operations many millions of times quicker than using classical machines, and this particular aspect makes it particularly suitable to computing tasks which demand vast amounts of computer processing. Machine learning (ML), or paradigm of allowing systems to recognise structures in the data, and then predict them using programs that are designed to find patterns in the data, is, on the other hand, the lifeblood of contemporary artificial intelligence. With the added advantage of speed and computational capabilities of quantum computing combined with the capabilities of machine learning, QML will be used to transform many areas such as cybersecurity to provide superior abilities that fall well outside the scope of classical systems (Cerezo et al., 2021).

Quantum computing has unique properties that make it possible to address complex tasks using quantum machine learning because of superposition and entanglement. Quantum machine learning not merely speed up the computation of massive, but also allow the formulation of more advanced algorithms solving previously unsolvable problems. Such a synergy between the quantum computing and machine learning will transform the methodologies of cybersecurity, providing a means to detect threats at a faster rate, encrypt them more effectively, and create robust cloud environments (Bowles, Ahmed, & Schuld, 2024). Allowing QML to be incorporated in cloud systems, this is a potent option to boost the safety of the data, and it will enable people to select options to keep their sensitive information safe, without which the digital world may be perforated (Payares & Martinez-Santos, 2021).

1.3 Quantum ML and applications in Security in the cloud

Among the possible applications of QML, it can be touted as one of the tools that will speed up the algorithms of machine learning, especially as applied to security on the cloud. Another fundamental field of QML research is quantum-accelerated linear algebra, because quantum computers can compute matrix operations and transformations of vectors by far moreverbidnice billion times faster than classical computers (Gouveia & Correia, 2020). Linear algebra operations are a crucial part of machine learning work such as principal component analysis (PCA), k-means clustering and support vector machines (SVMs). The fact that quantum computing can speed up these operations makes machine learning systems handle more data, which is essential in the case of preventing threats and anomaly tracking in real-time in cloud systems.

New quantum algorithms, like quantum support vector machines (QSVM) and quantum neural networks (QNN) can also be found in QML and they can surpass their classical equivalents. An example is that QSVMs exploit quantum parallelism allowing the rapid classification of data and pattern recognition. Such quantum-enhanced models are especially applicable in systems involving cybersecurity, where the data such as network traffic, user, and system logs require real-time analysis in order to recognize and address threats (Adcock et al., 2015). In addition, quantum-inspired machine learning approaches use quantum concepts to solve machine learning problems related to classical methods without resorting to entire quantum hardware, which to some extent creates a convenient solution to incorporating quantum features to a cloud-based system (Bonguet & Bellaiche, 2017).

1.4 Quantum machine learning methods to Cybersecurity

The implementation of QML in cybersecurity can be viewed as multiple versions of doing so, and each of them has its specific benefits:

a) **Quantum Machine Learning Algorithms:** The quantum Machine learning algorithms use classical machine learning algorithm and translate them into quantum ones. It can be seen in the form of implementing a quantum analogue to support vector machines or QSVMs, quantum neural networks or QNNs which are optimally used to leverage the computing power of quantum computers to present a classification on data and to make decisions (Farhi & Neven, 2018). These quantum machine learning models are much faster and efficient in processing the data especially when it comes to security on clouds, where in real-time, the time spent is of prime importance in identifying any threat.
b) **Quantum-Inspired Machine Learning:** Quantum-inspired machine learning stores ideas of quantum computing in order to improve classical machine learning algorithms. In one such way, quantum-inspired classifiers (namely Helstrom Quantum Centroid (HQC) and quantum-inspired binary classifiers (QIBC)) utilize quantum concepts to enhance the precision of classification problems in cybersecurity (Feng, Cai, & Zhou, 2019). These methods create an avenue of utilizing quantum concepts into classical systems without their need to have complete quantum infrastructure and therefore they can find their place in real life cybersecurity use cases.
c) **Hybrid Quantum-Classical Machine Learning:** The hybrid method uses both quantum algorithms and classical machine learning strategies so that the advantage of both technologies can be used. Hybrid quantum-classical models Hybrid quantum-classical models Hybrid quantum-classical models are expected to enhance the capabilities of classical algorithms by providing quantum processing capabilities, examples of which include quantum-enhanced support vector machines (QSVM) and quantum classifiers. Such of a kind makes it possible to find solutions to difficult classification and regression tasks typical of cybersecurity, including intrusion detection systems and anomaly detection (Dallaire-Demers et al., 2018).

1.5 Quantum Machine Learning on Cloud Security

Quantum machine learning has the potential of transforming cloud-based security into an entirely new concept by solving some of the most critical issues affecting the area of digital data protection. Cryptography is one of the greatest fields of influence. Quantum computing is about to crack most of the encryption mechanisms that the digital world has used to securely transfer information in the past decades. Currently, the widespread encryption of cloud communication over the public key system can be made useless by defining quantum algorithms that can address mathematical issues impossible to present conventional computers (Payares & Martinez-Santos, 2021). As an example, quantum algorithms like Shor algorithm can factor large numbers exponentially faster than only classical ones, so there is a real threat to RSA encryption.

Quantum machine learning can be used to solve these threats by offering innovative cryptographic techniques that are intrinsically unsuited to quantum attacks. One of these is the Quantum Key Distribution (QKD) that applies the concepts of quantum mechanics to guarantee secure communication by ensuring that the key can not be intercepted by the eavesdropper without his/her detection. The quantum-safe encryption strategies are crucial in protecting cloud-based communications and data in storage against cyber threats powered by quantum computers (Cerezo et al., 2021).

Quantum machine learning will also be essential in enhancing threat detection and anomaly tracing within the clouds as it continues to grow to maturity. QML has the capability of analyzing huge data at unbelievably fast rates so it is able to identify suspicious activity in real time and therefore an organization is able to recognize possible security breaches arising before they can reach an advanced stage. As an illustration, we can say that the network traffic and the activity of users can be monitored with the help of quantum-inspired models of machine learning, offering an effective method to identify malicious actions like distributed denial-of-service (DDoS) attack or a ransomware infection (Brewer, 2016).

1.6 Quantum Machine Learning Security in Cloud-Based Security

There are many applications of quantum machine learning in cloud-based security with all of them having a positive effect on the capability of protecting sensitive information and even enhancing the threat detection capabilities:

a) **Improvement of the Decryption and Encryption Process:** To protect the security and integrity of data in the cloud environment, quantum-safe cryptographic algorithms play a crucial role in the process. The quantum machine learning can be used to speed up the process of coming up with quantum-resistant cryptography like Quantum Key Distribution (QKD) which is resistant to the quantum attack. Such encryption algorithms are going to make sure that critical information is encrypted, even in conditions of quantum-capable attackers (Bharti et al., 2022).

b) **Threat detection and anomaly tracking:** Real-time anomaly tracking and threat detection is an exceptional aspect of the application of quantum machine learning to cybersecurity. Cloud systems allow to better analyze enormous amounts of data with the help of quantum-enhanced algorithms to detect suspect patterns and behaviors in order to respond quicker to possible security breaches in a case (Fredrikson, Jha, & Ristenpart, 2015). Parallel processing of the data with quantum machine learning allows identifying anomalies more efficiently and boosts system security as a whole.
c) **Quantum-Driven Authentication:** Authentication systems play an important role in the security of clouds and quantum machine learning is helping these systems to contribute by offering identity verification that is resistant to tampering using quantum entanglement. The quantum-powered authentication systems are more secure than the conventional ones, which minimizes the possibility of identity stealing and unauthorized access (Shah, Jhanjhi, & Brohi, 2024). These systems are an efficient solution to cloud-based services, as they use the principles of machine learning combined with the principles of quantum theory to create a secure and resilient environment.

Quantum machine learning is going to constitute a paradigm change in how we process and secure data in the cloud. Harnessing the processing speed and computing power of quantum computing combined with the highly advanced algorithms of machine learning, QML will become the disruption to cloud-based security bringing quicker encryption rates, expediency in detecting threats, and effective authentication models. It can only be foreseen that the further development of quantum technologies will bring about the use of quantum machine learning in cloud security, and this will become even more significant in terms of protection of digital information against ever-more advanced cyber threats that will emerge in future.

1.6 Challenges in Quantum Machine Learning for Cloud-Based Security

Despite the numerous benefits that can be obtained upon integrating Quantum Machine Learning (QML) in cloud-based security services, a couple of issues should be addressed. The way forward has a number of steps, such as designing quantum-safe communication protocols, addressing the gap between research and commercial quantum hardware and improving expertise in quantum computing and machine learning. It involves breeding collaborations between machine learning experts, cybersecurity experts, and quantum researchers to make the strategic and collaborative embedding of quantum tech within existing cloud security systems.

The challenges should be overcome in order to achieve the full potential of the quantum-enhanced security systems.

a) Complicatedness: The field of quantum computing is very complicated and one has to possess certain knowledge and experience to work in that domain. Its complex nature may make cloud-based quantum computing services difficult to access, and difficult to use.
b) Scalability: Scalability is a problem in quantum as well as in conventional computing. Still, combining quantum computing and cloud computing has additional challenges. As an example, dynamic resource management and allocation have to be carried out and also low-latency and high communication between quantum and classical parts is required.
c) Security: There is a need to protect sensitive quantum algorithms and data against unauthorized access and interference following the joining of The quantum computing and cloud computing.
d) Interoperability: Since quantum computing advances, it may be required that different quantum computing systems and technologies can communicate with one another. This may be challenging because various programming models and technologies can be adopted on various platforms.
e) Price: The establishment of a quantum computing system and the acquisition of the necessary software and computer equipment may be demanding. Despite the possibility of using cloud computing being cheaper than the alternatives, there is the probability of some of these users finding the ability to access and utilize the cloud available quantum computing costs expensive.
f) Maturity: A good number of the technologies and applications in quantum computing are immature in terms of their development since the field is still young and is experiencing rapid development. Due to this fact, the process of predicting the development of quantum computing and its integration with cloud computing will be rather difficult.

Figure 2. Challenges in Quantum Machine Learning for Cloud-Based Security.

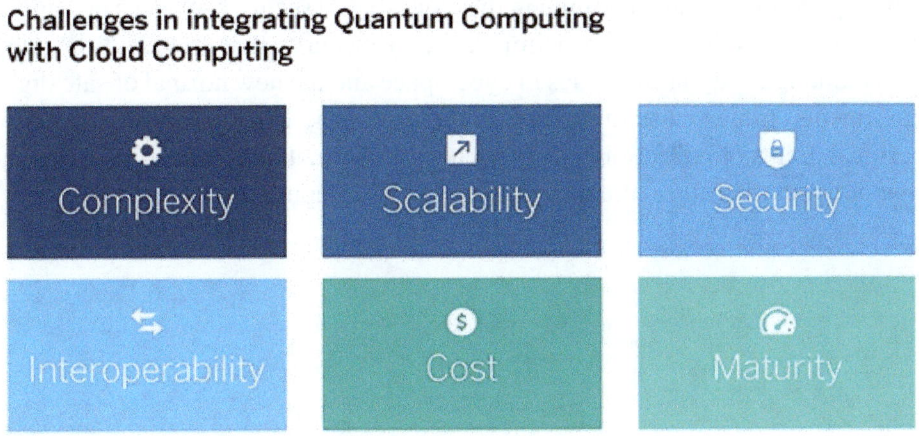

1.7 Future Directions for Quantum Machine Learning in Cloud-Based Security

The present level of Quantum Machine Learning (QML) in cloud-based security is only the tip of the iceberg where multiple future directions have promising possibilities. Further study on quantum-safe cryptographic algorithms, to create plausible scaled-quantum computing architecture, and to democratize quantum technologies must happen in the further development of QML in cloud security. With the maturing level of quantum computing technology, quantum-enhanced security applications will bring a surge of opportunities in cloud services. Democratizing quantum technologies will render quantum cloud computing clearer to the users and working with the power of quantum algorithms without a significant investment in on-premise quantum hardware will be achievable by organizations.

CONCLUSION

To sum up, the intersection of Quantum Machine Learning (QML) and cloud security is a revolutionary period of digital security. Aspiring to be revolutionary, Cloud-Enabled Quantum Machine Learning has its challenges. Complexity with error rates affecting quantum operations and effects of external events on quantum computers are some of the challenges sophisticated quantum hardware will have to face. This long chapter has given detailed overview of the possible uses of this innovative fusion, challenges, and possible direction of this breakthrough. Implementation of quantum machine learning in the cloud-based security services presents an opportunity to establish a new benchmark in cybersecurity that provides unmatched security among the dynamic issues in cyberspace and the new normal of safe digital environments. Ethics, sustainability and the use of the latest technologies will be the critical points of reflection during such a transformative era and will contribute to mapping out the future of a new golden age in computing.

REFERENCES

Adcock, J., Allen, E., Day, M., Frick, S., Hinchliff, J., Johnson, M., Morley-Short, S., Pallister, S., Price, A., & Stanisic, S. (2015). Advances in quantum machine learning. *Science*.

Anagolum, S., Alquraishi, A., Wang, Y., & Wang, H. (2024). Élivágar: Efficient quantum circuit search for classification. In *Proceedings of the 29th ACM International Conference on Architectural Support for Programming Languages and Operating Systems,* Volume 2 (pp. 336–353). DOI: 10.1145/3620665.3640354

Ash-Saki, A., Alam, M., & Ghosh, S. (2020). Analysis of crosstalk in NISQ devices and security implications in multi-programming regime. In *Proceedings of the ACM/IEEE International Symposium on Low Power Electronics and Design* (pp. 25–30). DOI: 10.1145/3370748.3406570

Ayanzadeh, R., Nguyen, H. N., & Proctor, T. J. (2023). Enigma: Privacy-preserving execution of QAOA on untrusted quantum computers. *arXiv preprint arXiv:2311.13546*.

Balewski, J., Mukherjee, M., Pérez, A. R., & Häffner, H. (2024). Quantum-parallel vectorized data encodings and computations on trapped-ion and transmon QPUs. *Scientific Reports, 14*(1), 3435. DOI: 10.1038/s41598-024-53720-x PMID: 38341454

Bharti, K., Cervera-Lierta, A., Haug, T., Pandey, A. K., Vishwanath, S., Kim, P. W., & Aspuru-Guzik, A. (2022). Noisy intermediate-scale quantum algorithms. *Reviews of Modern Physics, 94*(1), 015004. DOI: 10.1103/RevModPhys.94.015004

Bonguet, A., & Bellaiche, M. (2017). A survey of denial-of-service and distributed denial-of-service attacks and defenses in cloud computing. *Future Internet, 9*(3), 43. DOI: 10.3390/fi9030043

Bowles, J., Ahmed, S., & Schuld, M. (2024). Better than classical? The subtle art of benchmarking quantum machine learning models. *arXiv preprint arXiv:2403.07059*.

Brewer, R. (2016). Ransomware attacks: Detection, prevention, and cure. *Network Security, 2016*(9), 5–9. DOI: 10.1016/S1353-4858(16)30086-1

Cerezo, M., Sone, A., Volkoff, T., Cincio, L., & Coles, P. J. (2021). Variational quantum algorithms. *Nature Reviews. Physics, 3*(9), 625–644. DOI: 10.1038/s42254-021-00348-9

Dallaire-Demers, P.-L., Killoran, N., Qureshi, I. N., & Arrazola, J. M. (2018). Quantum generative adversarial networks. *Physical Review. A, 98*(1), 012324. DOI: 10.1103/PhysRevA.98.012324

Farhi, E., & Neven, H. (2018). Classification with quantum neural networks on near-term processors. *arXiv preprint arXiv:1802.06002*.

Feng, J., Cai, Q.-Z., & Zhou, Z.-H. (2019). Learning to confuse: Generating training-time adversarial data with auto-encoder. *Advances in Neural Information Processing Systems*, •••, 32.

Fredrikson, M., Jha, S., & Ristenpart, T. (2015). Model inversion attacks that exploit confidence information and basic countermeasures. In *Proceedings of the 22nd ACM SIGSAC Conference on Computer and Communications Security* (pp. 1322–1333). DOI: 10.1145/2810103.2813677

Gouveia, A., & Correia, M. (2020). Towards quantum-enhanced machine learning for network intrusion detection. In *2020 IEEE 19th International Symposium on Network Computing and Applications* (pp. 1–8). DOI: 10.1109/NCA51143.2020.9306691

Malhotra, Y. (2022, June). *How you can implement well-architected 'zero trust' hybrid-cloud computing beyond 'lift and shift': Cloud-enabled digital innovation at scale with infrastructure as code (IaC), DevSecOps, and MLOps. In 2022 New York State Cyber Security Conference*. Invited Presentations., https://its.ny.gov/2022-nyscsc

Payares, E., & Martinez-Santos, J. C. (2021). Quantum machine learning for intrusion detection of distributed denial of service attacks: A comparative overview. In *Proceedings of SPIE Quantum Computing, Communication, and Simulation* (p. 47). DOI: 10.1117/12.2593297

Shah, I. A., Jhanjhi, N., & Brohi, S. N. (2024). Proposing model for classification of malicious SQLi code using machine learning approach. In *2024 1st International Conference on Innovative Engineering Sciences and Technological Research (ICIESTR)* (pp. 1-6). DOI: 10.1109/ICIESTR60916.2024.10798230

Compilation of References

Abbasi, M., Shahraki, A., & Taherkordi, A. (2021). Deep Learning for Network Traffic Monitoring and Analysis (NTMA): A Survey. *Computer Communications*, *170*, 19–41. DOI: 10.1016/j.comcom.2021.01.021

Aboukadri, S., Ouaddah, A., & Mezrioui, A. (2024). Machine learning in identity and access Management Systems: Survey and deep dive. *Computers & Security*, *139*, 103729. DOI: 10.1016/j.cose.2024.103729

Adcock, J., Allen, E., Day, M., Frick, S., Hinchliff, J., Johnson, M., Morley-Short, S., Pallister, S., Price, A., & Stanisic, S. (2015). Advances in quantum machine learning. *Science*.

Agapito, G., & Cannataro, M. (2023). An overview on the challenges and limitations using cloud computing in healthcare corporations. *Big Data and Cognitive Computing*, *7*(2), 68. DOI: 10.3390/bdcc7020068

Agarkar, A. A., Karyakarte, M., Chavhan, G., Patil, M., Talware, R., & Kulkarni, L. (2024). Blockchain aware decentralized identity management and access control system. *Measurement. Sensors*, *31*, 101032. DOI: 10.1016/j.measen.2024.101032

Aghili, S. (2024). *Leveraging blockchain technology: Governance*. Risk, Compliance, Security, and Benevolent Use Cases. DOI: 10.1201/9781003462033

Ahamad, S. S. (2021). A novel NFC-based secure protocol for merchant transactions. *IEEE Access : Practical Innovations, Open Solutions*, *10*, 1905–1920. DOI: 10.1109/ACCESS.2021.3139065

Akhil. (2024, July 4). *How to Monetize your Mobile App: a guide for entrepreneurs*. Apptunix Blog. https://www.apptunix.com/blog/mobile-app-monetization/

Al-Abri, A., Madiseh, F. R., & Moghaddam, M. M. (2024). Exploring Learning-Oriented Assessment in Enhancing Students' lexical Fluency through MALL. *The Asia-Pacific Education Researcher*. Advance online publication. DOI: 10.1007/s40299-024-00832-7

Alajarmeh, N., & Ladner, R. E. (2024). Password managers use among individuals who are visually impaired: Awareness, adoption, and rejection. *International Journal of Human-Computer Interaction*, •••, 1–17. DOI: 10.1080/10447318.2024.2376356

Alassery, F., Ahmed, W., Sarraf, M., & Lawrence, V. (2015). Collision detection in wireless sensor networks through Pseudo-Coded ON-OFF pilot periods per packet: A novel Low-Complexity and Low-Power design technique. *Computer and Information Science*, *8*(3). Advance online publication. DOI: 10.5539/cis.v8n3p13

Aldughayfiq, B., Ashfaq, F., Jhanjhi, N. Z., & Humayun, M. (2023). YOLO-Based Deep Learning Model for Pressure Ulcer Detection and Classification. *Health Care*, *11*(9), 1222. DOI: 10.3390/healthcare11091222 PMID: 37174764

Alenizi, A., Mishra, S., & Baihan, A. (2024). Enhancing secure financial transactions through the synergy of blockchain and artificial intelligence. *Ain Shams Engineering Journal*, *15*(6), 102733. DOI: 10.1016/j.asej.2024.102733

Alex, S. A., Jhanjhi, N., Humayun, M., Ibrahim, A. O., & Abulfaraj, A. W. (2022). Deep LSTM Model for Diabetes Prediction with Class Balancing by SMOTE. *Electronics (Basel)*, *11*(17), 2737. DOI: 10.3390/electronics11172737

Alferidah, D. K., & Jhanjhi, N. (2020). Cybersecurity Impact over Bigdata and IoT Growth. *2020 International Conference on Computational Intelligence (ICCI)*. DOI: 10.1109/ICCI51257.2020.9247722

Alkinani, M. H., Almazroi, A. A., Jhanjhi, N., & Khan, N. A. (2021). 5G and IoT Based Reporting and Accident Detection (RAD) System to Deliver First Aid Box Using Unmanned Aerial Vehicle. *Sensors (Basel)*, *21*(20), 6905. DOI: 10.3390/s21206905 PMID: 34696118

Almrezeq, N., Humayun, M., Alruwaili, M., Alanazi, S., & Jhanjhi, N. Z. (2023). Cyber security attacks and challenges in Saudi Arabia during COVID-19. *International Journal of Computer Science & Network Security*, *23*(10), 179–187.

Almufareh, M. F., Jhanjhi, N., Khan, N. A., Almuayqil, S. N., Humayun, M., & Javed, D. (2024). BertSent: Transformer-Based Model for Sentiment Analysis of Penta-Class Tweet Classification. *IEEE Access : Practical Innovations, Open Solutions*, *12*, 196803–196817. DOI: 10.1109/ACCESS.2024.3515836

Alothman, B., Alibrahim, O., Alenezi, N., Alhashemi, A., Alhashemi, M., Almardasi, D., Khattab, O., Joumaa, C., & Khan, M. (2023). The Development of a Secure Online System to Protect Social Networking Platforms from Security Attacks. *Applied Sciences (Basel, Switzerland)*, *13*(21), 11731. DOI: 10.3390/app132111731

Al-Quayed, F., Javed, D., Jhanjhi, N. Z., Humayun, M., & Alnusairi, T. S. (2024). A Hybrid Transformer-Based Model for Optimizing Fake News Detection. *IEEE Access : Practical Innovations, Open Solutions*, *12*, 160822–160834. DOI: 10.1109/ACCESS.2024.3476432

Alrawili, R., AlQahtani, A. S., & Khan, M. K. (2024). Comprehensive survey: Biometric user authentication application, evaluation, and discussion. *Computers & Electrical Engineering*, *119*, 109485. DOI: 10.1016/j.compeleceng.2024.109485

Al-Shamaileh, O., & Sutcliffe, A. (2022). Why people choose Apps: An evaluation of the ecology and user experience of mobile applications. *International Journal of Human-Computer Studies*, *170*, 102965. DOI: 10.1016/j.ijhcs.2022.102965

Alshamrani, A., Myneni, S., Chowdhary, A., & Huang, D. (2019). A survey on Advanced persistent threats: Techniques, solutions, challenges, and research opportunities. *IEEE Communications Surveys and Tutorials*, *21*(2), 1851–1877. DOI: 10.1109/COMST.2019.2891891

Alsmadi, I. (2023). Identity Management. In *Springer eBooks* (pp. 313–329). DOI: 10.1007/978-3-031-21651-0_12

Alzaabi, F. R., & Mehmood, A. (2024). A Review of Recent Advances, Challenges and Opportunities in Malicious Insider Threat Detection using Machine Learning Methods. *IEEE Access : Practical Innovations, Open Solutions*, *1*, 30907–30927. Advance online publication. DOI: 10.1109/ACCESS.2024.3369906

Al-Zubaidie, M., & Jebbar, W. (2024). Transaction security and management of blockchain-based smart contracts in e-banking-employing microsegmentation and yellow saddle Goatfish. *Mesopotamian Journal of CyberSecurity*, *4*(2), 1–19. DOI: 10.58496/MJCS/2024/005

Amanat, A., Rizwan, M., Maple, C., Zikria, Y. B., Almadhor, A. S., & Kim, S. W. (2022). Blockchain and cloud computing-based secure electronic healthcare records storage and sharing. *Frontiers in Public Health*, *10*, 938707. DOI: 10.3389/fpubh.2022.938707 PMID: 35928494

Ammattikorkeakoulu, J. (2015). *Gamification of the sales process at a telecommunications company to improve the motivation of the salesforce*. Theseus. https://www.theseus.fi/handle/10024/133103

Anagolum, S., Alquraishi, A., Wang, Y., & Wang, H. (2024). Élivágar: Efficient quantum circuit search for classification. In *Proceedings of the 29th ACM International Conference on Architectural Support for Programming Languages and Operating Systems,* Volume 2 (pp. 336–353). DOI: 10.1145/3620665.3640354

Ananna, F. F., Nowreen, R., Jahwari, S. S. R. A., Costa, E. A., Angeline, L., & Sindiramutty, S. R. (2023). Analysing Influential factors in student academic achievement: Prediction modelling and insight. *International Journal of Emerging Multidisciplinaries Computer Science & Artificial Intelligence*, *2*(1). Advance online publication. DOI: 10.54938/ijemdcsai.2023.02.1.254

Andronache, G., & Aciobănitei, L. (2024). DAnke - An Application for Promoting a Healthy Lifestyle Using Gamification and Blockchain. *2024 IEEE 18th International Symposium on Applied Computational Intelligence and Informatics (SACI)*, 000365–000370. DOI: 10.1109/SACI60582.2024.10619826

Anvari, S. S., Hammer, J., & Wehbe, R. R. (2024). "More than just a game, it's an app that builds awareness around Mental Health": Mental Health Stigma Reduction Using Games for Change. *Proceedings of the ACM on Human-Computer Interaction, 8*(CHI PLAY), 1–30. DOI: 10.1145/3677090

Archibong, E. E., Stephen, B. U., & Asuquo, P. (2024). Analysis of cybersecurity vulnerabilities in mobile payment applications. *Archives of Advanced Engineering Science*. DOI: 10.47852/bonviewAAES42022595

Arenas, Á., Ray, G., Hidalgo, A., & Urueña, A. (2023). How to keep your information secure? Toward a better understanding of users security behavior. *Technological Forecasting and Social Change*, *198*, 123028. DOI: 10.1016/j.techfore.2023.123028

Arora, A., Garg, S., & Peddoju, S. K. (2014). Malware Detection Using Network Traffic Analysis in Android Based Mobile Devices. *2014 Eighth International Conference on Next Generation Mobile Apps, Services and Technologies*, 66–71. DOI: 10.1109/NGMAST.2014.57

Arroteia, N. F. P. (2024). Unveiling the invisible: Soft skills recognition from social media. *Company Games & Business Simulation Academic Journal*, *4*(1), 69. DOI: 10.54988/cg.2024.1.1573

Asaad, S. M., & Maghdid, H. S. (2022). A comprehensive review of indoor/outdoor localization solutions in IoT era: Research challenges and future perspectives. *Computer Networks*, *212*, 109041. DOI: 10.1016/j.comnet.2022.109041

Asatiani, A., Hakkarainen, T., Paaso, K., & Penttinen, E. (2023). Security by envelopment – a novel approach to data-security-oriented configuration of lightweight-automation systems. *European Journal of Information Systems*, •••, 1–23. DOI: 10.1080/0960085X.2023.2217362

Ashok, P., & Hallur, G. (2024). Navigating the Infotainment Landscape: Status and Regulations in the Digital Age. *2024 International Conference on Advances in Data Engineering and Intelligent Computing Systems (ADICS)*, 1–6. DOI: 10.1109/ADICS58448.2024.10533560

Ashraf, H., Jhanjhi, N. Z., Ihsan, U., Khan, N. A., & Humayun, M. (2025). Prediction of health indicators based on exercise, sleep, and water intake: Data-based machine learning approach. In *AI Techniques for Securing Medical and Business Practices* (pp. 291–310). IGI Global.

Ashraf, H., Khan, F., Ihsan, U., Al-Quayed, F., Jhanjhi, N. Z., & Humayun, M. (2023). MABPD: Mobile Agent-Based Prevention and Black Hole Attack Detection in Wireless Sensor Networks. *2023 International Conference on Business Analytics for Technology and Security (ICBATS)*, 1–11. DOI: 10.1109/ICBATS57792.2023.10111277

Ash-Saki, A., Alam, M., & Ghosh, S. (2020). Analysis of crosstalk in NISQ devices and security implications in multi-programming regime. In *Proceedings of the ACM/IEEE International Symposium on Low Power Electronics and Design* (pp. 25–30). DOI: 10.1145/3370748.3406570

Audestad, J. A., & Telematikk, N. T. U. F. F. I. M. O. E. I. F. (2014). *The future of the mobile application market*. https://ntnuopen.ntnu.no/ntnu-xmlui/handle/11250/262967

Ayanzadeh, R., Nguyen, H. N., & Proctor, T. J. (2023). Enigma: Privacy-preserving execution of QAOA on untrusted quantum computers. *arXiv preprint arXiv:2311.13546*.

Azam, H., Dulloo, M. I., Majeed, M. H., Wan, J. P. H., Xin, L. T., & Sindiramutty, S. R. (2023). Cybercrime Unmasked: Investigating cases and digital evidence. *International Journal of Emerging Multidisciplinaries Computer Science & Artificial Intelligence*, *2*(1). Advance online publication. DOI: 10.54938/ijemdcsai.2023.02.1.255

Azam, H., Dulloo, M. I., Majeed, M. H., Wan, J. P. H., Xin, L. T., Tajwar, M. A., & Sindiramutty, S. R. (2023). Defending the digital Frontier: IDPS and the battle against Cyber threat. *International Journal of Emerging Multidisciplinaries Computer Science & Artificial Intelligence*, *2*(1). Advance online publication. DOI: 10.54938/ijemdcsai.2023.02.1.253

Azam, H., Tajwar, M. A., Mayhialagan, S., Davis, A. J., Yik, C. J., Ali, D., & Sindiramutty, S. R. (2023). Innovations in Security: A study of cloud Computing and IoT. *International Journal of Emerging Multidisciplinaries Computer Science & Artificial Intelligence*, *2*(1). Advance online publication. DOI: 10.54938/ijemdcsai.2023.02.1.252

Azam, H., Tan, M., Pin, L. T., Syahmi, M. A., Qian, A. L. W., Jingyan, H., Uddin, M. F., & Sindiramutty, S. R. (2023). Wireless Technology Security and Privacy: A Comprehensive Study. *Preprints.org*. DOI: 10.20944/preprints202311.0664.v1

Babbar, H., Rani, S., Masud, M., Verma, S., Anand, D., & Jhanjhi, N. (2021). Load balancing algorithm for migrating switches in software-defined vehicular networks. *Computers, Materials & Continua/Computers. Computers, Materials & Continua*, *67*(1), 1301–1316. DOI: 10.32604/cmc.2021.014627

Bachoo, N., & Nagowah, L. (2024). GAMIFIE: A gamified framework as a strategy to increase motivation and engagement among employees in IT industries. In *Signals and communication technology* (pp. 95–118). DOI: 10.1007/978-3-031-57880-9_5

Baek, S., Ahn, J., & Kim, D. (2023). Future Business Model for Mobile Cloud Gaming: The Case of South Korea and Implications. *IEEE Communications Magazine*, *61*(7), 68–73. DOI: 10.1109/MCOM.001.2200374

Bag, A., Patranabis, S., & Mukhopadhyay, D. (2023). CAMISE: Content Addressable Memory-Integrated Searchable Encryption. *IEEE Transactions on Circuits and Systems. I, Regular Papers*, *70*(8), 3254–3267. DOI: 10.1109/TCSI.2023.3279853

Baidya, R., Kumar, D., & Livvarcin, D. O. (2024). Gamified communication in social media for brand advocacy of Start-Ups. In *Routledge eBooks* (pp. 170–191). DOI: 10.4324/9781032694238-14

Balasubramaniam, S., Sankar, K. S., Malar, C. B., Rajasekaran, M. P., Arunkumar, N., & Thilagaraj, M. (2023). Detecting the type of attack in healthcare cloud computing system using termite deep belief neural network (TDBNN) for future smart cities. *Journal of Testing and Evaluation*, *51*(3), 1421–1437. DOI: 10.1520/JTE20220041

Balewski, J., Mukherjee, M., Pérez, A. R., & Häffner, H. (2024). Quantum-parallel vectorized data encodings and computations on trapped-ion and transmon QPUs. *Scientific Reports*, *14*(1), 3435. DOI: 10.1038/s41598-024-53720-x PMID: 38341454

Bapna, R., & Ghose, A. (2024). *Thrive: Maximizing Well-Being in the Age of AI*. MIT Press. DOI: 10.7551/mitpress/15438.001.0001

Barshandeh, S., Masdari, M., Dhiman, G., Hosseini, V., & Singh, K. K. (2021). A range-free localization algorithm for IoT networks. *International Journal of Intelligent Systems, 37*(8), 10336–10379. DOI: 10.1002/int.22720

Batra, I., Luhach, A. K., & Pathak, N. (2016, March). Research and analysis of lightweight cryptographic solutions for internet of things. In *Proceedings of the second international conference on information and communication technology for competitive strategies* (pp. 1-5).

Bécue, I., Praça, I., & Gama, J. (2021). Artificial intelligence, cyber-threats and Industry 4.0: Challenges and opportunities. *Artificial Intelligence Review, 54*(5), 3849–3886. DOI: 10.1007/s10462-020-09942-2

Bellavista-Parent, V., Torres-Sospedra, J., & Pérez-Navarro, A. (2022). Comprehensive analysis of applied machine learning in indoor positioning based on Wi-Fi: An extended systematic review. *Sensors (Basel), 22*(12), 4622. DOI: 10.3390/s22124622 PMID: 35746404

Bergauer, A. (2024a). *Run it like a business: Strategies for Arts Organizations to Increase Audiences, Remain Relevant, and Multiply Money—Without Losing the Art*. BenBella Books.

Bernstein, E., & Vazirani, U. (1993). Quantum complexity theory. In *Proceedings of the twenty-fifth annual ACM symposium on Theory of computing - STOC '93* (pp. 11–20). New York, NY, USA: ACM Press. DOI: 10.1145/167088.167097

Bharti, K., Cervera-Lierta, A., Haug, T., Pandey, A. K., Vishwanath, S., Kim, P. W., & Aspuru-Guzik, A. (2022). Noisy intermediate-scale quantum algorithms. *Reviews of Modern Physics, 94*(1), 015004. DOI: 10.1103/RevModPhys.94.015004

Bhatt, P., Sethi, A., Tasgaonkar, V., Shroff, J., Pendharkar, I., Desai, A., Sinha, P., Deshpande, A., Joshi, G., Rahate, A., Jain, P., Walambe, R., Kotecha, K., & Jain, N. K. (2023). Machine learning for cognitive behavioral analysis: Datasets, methods, paradigms, and research directions. *Brain Informatics, 10*(1), 18. Advance online publication. DOI: 10.1186/s40708-023-00196-6 PMID: 37524933

Bizzi, L. (2023). Why to gamify performance management? Consequences of user engagement in gamification. *Information & Management, 60*(3), 103762. DOI: 10.1016/j.im.2023.103762

Bloom, P. (2023). *The human mind: A Brief Tour of Everything We Know*. Random House.

Bojjagani, S., Seelam, N. R., Sharma, N. K., Uyyala, R., Akuri, S. R. C. M., & Maurya, A. K. (2023). The use of iot-based wearable devices to ensure secure lightweight payments in fintech applications. *Journal of King Saud University. Computer and Information Sciences*, *35*(9), 101785. DOI: 10.1016/j.jksuci.2023.101785

Bolat, Y. I., & Taş, N. (2022). A meta-analysis on the effect of gamified-assessment tools' on academic achievement in formal educational settings. *Education and Information Technologies*, *28*(5), 5011–5039. DOI: 10.1007/s10639-022-11411-y

Bonguet, A., & Bellaiche, M. (2017). A survey of denial-of-service and distributed denial-of-service attacks and defenses in cloud computing. *Future Internet*, *9*(3), 43. DOI: 10.3390/fi9030043

Boudlal, H., Serrhini, M., & Tahiri, A. (2022, May). Cloud computing for healthcare services: Technology, application, and architecture. In *2022 11th International Symposium on Signal, Image, Video and Communications (ISIVC)* (pp. 1-5). IEEE.

Boudreau, K. J., Jeppesen, L. B., & Miric, M. (2023). Free(mium) strategies for digital goods. In *Edward Elgar Publishing eBooks* (pp. 126–141). DOI: 10.4337/9781800378902.00012

Bowles, J., Ahmed, S., & Schuld, M. (2024). Better than classical? The subtle art of benchmarking quantum machine learning models. *arXiv preprint arXiv:2403.07059.*

Brewer, R. (2016). Ransomware attacks: Detection, prevention, and cure. *Network Security*, *2016*(9), 5–9. DOI: 10.1016/S1353-4858(16)30086-1

Brohi, S. N., Jhanjhi, N., Brohi, N. N., & Brohi, M. N. (2020). Key Applications of State-of-the-Art Technologies to Mitigate and Eliminate COVID-19.pdf. *TECHRxiv*. DOI: 10.36227/techrxiv.12115596.v1

Brown, M., Bamberger, P., Bliese, P. D., & Shields, J. (2023). Fairness uncertainty and pay information exchange: Why and when employees disclose bonus pay to pay information websites. *Journal of Organizational Behavior*, *44*(9), 1362–1379. DOI: 10.1002/job.2739

Bucko, A., Vishi, K., Krasniqi, B., & Rexha, B. (2023). Enhancing JWT authentication and authorization in web applications based on user behavior history. *Computers*, *12*(4), 78. DOI: 10.3390/computers12040078

Burghoorn, F., Scheres, A., Monterosso, J., Guo, M., Luo, S., Roelofs, K., & Figner, B. (2024). Pavlovian impatience: The anticipation of immediate rewards increases approach behaviour. *Cognitive, Affective & Behavioral Neuroscience*. Advance online publication. DOI: 10.3758/s13415-024-01236-2 PMID: 39467981

Busch, P., Heinonen, T., & Lahti, P. (2007, November). T. HEINONEN, and P. LAHTI, "Heisenberg's uncertainty principle,". *Physics Reports*, *452*(6), 155–176. DOI: 10.1016/j.physrep.2007.05.006

Cacciotto, F., Fulcini, T., Coppola, R., & Ardito, L. (2021, April). A metric framework for the gamification of web and mobile gui testing. In *2021 IEEE International Conference on Software Testing, Verification and Validation Workshops (ICSTW)* (pp. 126-129). IEEE. DOI: 10.1109/ICSTW52544.2021.00032

Cahya, A., & Indriana, D. (2023, August 30). *Transforming middle School mathematics education: The impact of gamification on student interest and engagement.* https://iocscience.org/ejournal/index.php/geneus/article/view/4749

Cao, Y., Gong, S., Wang, Y., Zheng, Q., & Wang, Z. (2022). How to provide competitors in educational gamification: The roles of competitor level and autonomous choice. *Computers in Human Behavior*, *138*, 107477. DOI: 10.1016/j.chb.2022.107477

Casolare, R., De Dominicis, C., Iadarola, G., Martinelli, F., Mercaldo, F., & Santone, A. (2021). Dynamic mobile malware detection through system call-based image representation. *Journal of Wireless Mobile Networks, Ubiquitous Computing and Dependable Applications*, *12*(1), 44–63. DOI: 10.22667/JOWUA.2021.03.31.044

Cerezo, M., Sone, A., Volkoff, T., Cincio, L., & Coles, P. J. (2021). Variational quantum algorithms. *Nature Reviews. Physics*, *3*(9), 625–644. DOI: 10.1038/s42254-021-00348-9

Chakraborty, S., Jackson, C., Frazier, M., & Clark, K. (2024). A Study on Password Protection and Encryption in the era of Cyber Attacks. *SoutheastCon*, *2024*, 460–465. Advance online publication. DOI: 10.1109/SoutheastCon52093.2024.10500214

Chan, G., Arya, A., Orji, R., Zhao, Z., & Whitehead, A. (2023). Increasing motivation in social exercise games: Personalising gamification elements to player type. *Behaviour & Information Technology*, •••, 1–31. DOI: 10.1080/0144929X.2023.2255293

Chan, G., Nwagu, C., Odenigbo, I., Alslaity, A., & Orji, R. (2024). The Shape of Mobile Health: A Systematic review of health visualization on mobile devices. *International Journal of Human-Computer Interaction*, •••, 1–19. DOI: 10.1080/10447318.2024.2313282

Chapman, J. R., Kohler, T. B., & Gedeborg, S. (2023). So, why do students perform better in gamified courses? Understanding motivational styles in educational gamification. *Journal of Educational Computing Research*, *61*(5), 927–950. DOI: 10.1177/07356331221127635

Châtel, B. D. L., Janssen, J. H. M., Peeters, G. M. E. E., Corten, R., Tieben, R., Deen, M., Hendriks, E. J. M., & Rikkert, M. G. M. O. (2024). Social Gaming to Decrease loneliness in Older Adults: Recruitment Challenges and attrition Analysis in a digital Mixed Methods Feasibility study. *JMIR Serious Games*, *12*, e52640. DOI: 10.2196/52640 PMID: 39412863

Chauhan, R. P. S., Sonker, S. K., Kaur, M., Sharma, C., Singh, R., & Singh, R. (2024, March). Optimizing IoT Threat Mitigation with Artificial Intelligence in Banking: A Multi-Objective Approach. In *2024 2nd International Conference on Disruptive Technologies (ICDT)* (pp. 296-301). IEEE.

Chen, C., & Huang, K. (2024). The interplay of rewards and competition in digital game-based learning: Effects on intrinsic motivation, game performance and behaviors, and computational thinking. *Learning and Instruction*, *94*, 101995. DOI: 10.1016/j.learninstruc.2024.101995

Cheng, M., & Zhang, S. (2024). Reputation Burning: Analyzing the impact of brand sponsorship on social influencers. *Management Science*, mnsc.2023.00193. Advance online publication. DOI: 10.1287/mnsc.2023.00193

Chen, L., Chen, L., Jordan, S., Liu, Y. K., Moody, D., Peralta, R., & Smith-Tone, D. (2016). *Report on post-quantum cryptography* (Vol. 12). US Department of Commerce, National Institute of Standards and Technology.

Chen, Y. S., Hsu, C. S., & Huang, C. Y. (2021). A semi-supervised transfer learning with grid segmentation for outdoor localization over LoRaWANs. *Sensors (Basel)*, *21*(8), 2640. DOI: 10.3390/s21082640 PMID: 33918695

Chen, Y., Woodward, J., Bista, D., Zhang, X., Singh, I., Obajemu, O., Shankar, M. N., Ross, K. M., Ruiz, J., & Anthony, L. (2024). *Investigating Contextual Notifications to Drive Self-Monitoring in mHealth Apps for Weight Maintenance*. ACM., DOI: 10.1145/3613904.3641993

Chesti, I. A., Humayun, M., Sama, N. U., & Jhanjhi, N. (2020). Evolution, Mitigation, and Prevention of Ransomware. *2020 2nd International Conference on Computer and Information Sciences (ICCIS)*. DOI: 10.1109/ICCIS49240.2020.9257708

Chiu, N. P. C., Rupitan, M. J., King, K. K., Sharbini, H., Borhan, N. H., & Yeo, A. (2024). Effects of Gamification in space Exploration learning. *Journal of Advanced Research in Applied Sciences and Engineering Technology*, 265–274. DOI: 10.37934/araset.59.1.265274

Choksy, P., Chaurasia, A., Rao, U. P., & Kumar, S. (2023). Attribute based access control (ABAC) scheme with a fully flexible delegation mechanism for IoT healthcare. *Peer-to-Peer Networking and Applications*, *16*(3), 1445–1467. DOI: 10.1007/s12083-023-01486-w

Chowdhury, A. T. (n.d.). *The Impact Of5G Network On Industry 4.0*. DOI: 10.13140/RG.2.2.25308.16001

Ch, R., Batra, I., & Malik, A. (2021). Combining blockchain multi authority and botnet to create a hybrid adaptive crypto cloud framework. In *Proceedings of the 2021 International Conference on Computing Sciences (ICCS 2021)* (pp. 101–106). DOI: 10.1109/ICCS54944.2021.00028

Christopoulos, A., & Mystakidis, S. (2023). Gamification in education. *Encyclopedia*, *3*(4), 1223–1243. DOI: 10.3390/encyclopedia3040089

Chukwunweike, N. J. N., Yussuf, N. M., Okusi, N. O., Bakare, N. T. O., & Abisola, N. J. (2024). The role of deep learning in ensuring privacy integrity and security: Applications in AI-driven cybersecurity solutions. *World Journal of Advanced Research and Reviews*, *23*(2), 1778–1790. DOI: 10.30574/wjarr.2024.23.2.2550

Cigdem, H., Ozturk, M., Karabacak, Y., Atik, N., Gürkan, S., & Aldemir, M. H. (2024). Unlocking student engagement and achievement: The impact of leaderboard gamification in online formative assessment for engineering education. *Education and Information Technologies*, *29*(18), 24835–24860. Advance online publication. DOI: 10.1007/s10639-024-12845-2

Dahl, H. (2024). *How to Create Value in Social Media: a Postphenomenological Perspective : The technological mediation of social media between users and business - the effect on value of human factors engineering and customer engagement seen from a postphenomenological view*. DIVA. https://www.diva-portal.org/smash/record.jsf?pid=diva2%3A1882279&dswid=8584

Dallaire-Demers, P.-L., Killoran, N., Qureshi, I. N., & Arrazola, J. M. (2018). Quantum generative adversarial networks. *Physical Review. A*, *98*(1), 012324. DOI: 10.1103/PhysRevA.98.012324

Dastane, O., Fandos-Roig, J. C., & Sánchez-García, J. (2023). It's free! Still, would I learn? Unearthing perceived value of education apps for better entrepreneurial decisions. *Management Decision*. Advance online publication. DOI: 10.1108/MD-09-2022-1292

Davidavičienė, V., Raudeliūnienė, J., & Viršilaitė, R. (2021). Evaluation of user experience in augmented reality mobile applications. *Journal of Business Economics and Management*, *22*(2), 467–481. DOI: 10.3846/jbem.2020.13999

Delfianto, H. F., & Chandra, Y. U. (2023). The Impact of Key Factors on Subscription Service Adoption and Retention in Indonesia. *2023 International Conference on Computer Science, Information Technology and Engineering (ICCoSITE)*, 909–914. DOI: 10.1109/ICCoSITE57641.2023.10127749

Demszky, D., Liu, J., Hill, H. C., Jurafsky, D., & Piech, C. (2023). Can automated feedback improve teachers' uptake of student ideas? Evidence from a randomized controlled trial in a Large-Scale online course. *Educational Evaluation and Policy Analysis*, *46*(3), 483–505. DOI: 10.3102/01623737231169270

Denden, M., Tlili, A., Chen, N., Abed, M., Jemni, M., & Essalmi, F. (2022). The role of learners' characteristics in educational gamification systems: A systematic meta-review of the literature. *Interactive Learning Environments*, •••, 1–23. DOI: 10.1080/10494820.2022.2098777

Devi, P. J., Dutta, M. S., Damerakonda, M., Gutti, D. N., & Domakuntla, S. K. (2023). One-Time QR-Code for Fake Product Identification. *2023 5th International Conference on Inventive Research in Computing Applications (ICIRCA)*, 583–588. DOI: 10.1109/ICIRCA57980.2023.10220784

Dinata, H., Suryani, E., & Purnomo, J. D. T. (2024). Exploring the key success strategy for mobile game developers: A distinguish between premium and freemium versions. *Edelweiss Applied Science and Technology*, *8*(6), 1096–1108. DOI: 10.55214/25768484.v8i6.2213

Djaruma, H., Wirawan, Z. W., Gunawan, A. S., & Saputra, K. E. (2023). Systematic Literature Review of Monetization in Mobile App: Strategies, Trends, Challenges, and Best Practices. *2023 IEEE International Conference on Communication, Networks and Satellite (COMNETSAT)*. DOI: 10.1109/COMNETSAT59769.2023.10420748

Dogra, V., Singh, A., Verma, S., Kavita, N., Jhanjhi, N. Z., & Talib, M. N. (2021). Analyzing DistilBERT for Sentiment Classification of Banking Financial News. In *Lecture notes in networks and systems* (pp. 501–510). DOI: 10.1007/978-981-16-3153-5_53

Domoney, C. (2024). *Defending APIs: Uncover advanced defence techniques to craft secure application programming interfaces*. Packt Publishing Ltd.

Do, N., Jin, T., Priest, R., Meredith, L. N., & Landers, R. N. (2024). A longitudinal quasi-experiment of leaderboard effectiveness on learner behaviors and course performance. *Learning and Individual Differences*, *116*, 102572. DOI: 10.1016/j.lindif.2024.102572

Dong, B., Zhuang, M., Fang, E., & Huang, M. (2023). Tales of two channels: Digital advertising performance between AI recommendation and user subscription channels. *Journal of Marketing*. Advance online publication. DOI: 10.1177/00222429231190021

Dou, F., Lu, J., Xu, T., Huang, C. H., & Bi, J. (2021). A bisection reinforcement learning approach to 3-D indoor localization. *IEEE Internet of Things Journal*, *8*(8), 6519–6535. DOI: 10.1109/JIOT.2020.3041204

Dowd, M., McDonald, J., & Schuh, J. (2006). *The art of software security assessment: Identifying and Preventing Software Vulnerabilities*. Pearson Education.

Drott, E. (2024). Surveillance of and through Music. In *Surveillance of and through Music* (pp. 273–289). DOI: 10.1007/978-3-031-64013-1_17

Dudzinskaite, U., Correia, R., Venciute, D., & Fontes, R. (2024). Increasing customer engagement in digital marketing campaigns in a time of AI. In *Advances in marketing, customer relationship management, and e-services book series* (pp. 48–80). DOI: 10.4018/979-8-3693-2153-9.ch003

Dzidzah, E., Owusu Kwateng, K., & Asante, B. K. (2020). Security behaviour of mobile financial service users. *Information and Computer Security*, *28*(5), 719–741. DOI: 10.1108/ICS-02-2020-0021

Ebadi, S., Rasouli, R., & Mohamadi, M. (2021). Exploring EFL learners' perspectives on using Kahoot as a game-based student response system. *Interactive Learning Environments*, *31*(4), 2338–2350. DOI: 10.1080/10494820.2021.1881798

Echauri, G. (2023). Digital subscribers: Between freedom and constraint. *Media Culture & Society*, *45*(6), 1175–1191. DOI: 10.1177/01634437231159554

Eklund, A., Nygårds, A., & Smeds, D. (2018). *GeoFort: Mobile Game for Motivating Physical Activity using Gamification and Augmented Reality*. DIVA. https://www.diva-portal.org/smash/record.jsf?pid=diva2%3A1231096&dswid=-9383

Elhadad, A., Alanazi, F., Taloba, A. I., & Abozeid, A. (2022). Fog computing service in the healthcare monitoring system for managing the real-time notification. *Journal of Healthcare Engineering*, *2022*, 2022. DOI: 10.1155/2022/5337733 PMID: 35340260

Enache, A., Friberg, R., & Wiklander, M. (2023). Demand for in-app purchases in mobile apps—A difference-in-difference approach. *International Journal of Industrial Organization*, *88*, 102945. DOI: 10.1016/j.ijindorg.2023.102945

Etchells, P. (2024). *Unlocked: The Real Science of Screen Time (and how to spend it better)*. Hachette UK.

Faganel, A., Pačarić, F., & Rižnar, I. (2024). The impact of gamification on Slovenian consumers' online shopping. *Administrative Sciences*, *14*(5), 86. DOI: 10.3390/admsci14050086

Faizah, A., Hardian, A. F. A., Nandini, R. D., Handayani, P. W., & Harahap, N. C. (2024). The influence of Incentive-Based Mobile Fitness apps on Users' continuance Intention with Gender Moderation Effects: Quantitative and Qualitative study. *JMIR Human Factors*, *11*, e50957. DOI: 10.2196/50957 PMID: 38837199

Familoni, N. B. T., & Onyebuchi, N. N. C. (2024). AUGMENTED AND VIRTUAL REALITY IN U.S. EDUCATION: A REVIEW: ANALYZING THE IMPACT, EFFECTIVENESS, AND FUTURE PROSPECTS OF AR/VR TOOLS IN ENHANCING LEARNING EXPERIENCES. *International Journal of Applied Research in Social Sciences*, *6*(4), 642–663. DOI: 10.51594/ijarss.v6i4.1043

Farahsari, P. S., Farahzadi, A., Rezazadeh, J., & Bagheri, A. (2022). A survey on indoor positioning systems for IoT-based applications. *IEEE Internet of Things Journal*, *9*(10), 7680–7699. DOI: 10.1109/JIOT.2022.3149048

Fard Moshiri, P., Shahbazian, R., Nabati, M., & Ghorashi, S. A. (2021). A CSI-based human activity recognition using deep learning. *Sensors (Basel)*, *21*(21), 7225. DOI: 10.3390/s21217225 PMID: 34770532

Farhi, E., & Neven, H. (2018). Classification with quantum neural networks on near-term processors. *arXiv preprint arXiv:1802.06002*.

Faria, A. M., Dias, Á., Pereira, L. F., d, R. L., Costa, , & Gonçalves, R. (2024). Investigating factors that affect willingness to pay: An analysis on freemium social media apps. *International Journal of Advanced Media and Communication*, *8*(1), 12–52. DOI: 10.1504/IJAMC.2024.140638

Fatima-Tuz-Zahra, N., Jhanjhi, N., Brohi, S. N., Malik, N. A., & Humayun, M. (2020). Proposing a Hybrid RPL Protocol for Rank and Wormhole Attack Mitigation using Machine Learning. *2020 2nd International Conference on Computer and Information Sciences (ICCIS)*. DOI: 10.1109/ICCIS49240.2020.9257607

Feng, J., Cai, Q.-Z., & Zhou, Z.-H. (2019). Learning to confuse: Generating training-time adversarial data with auto-encoder. *Advances in Neural Information Processing Systems*, •••, 32.

Fernando, P. (2024). Exploring Perceptions and Habits of Sri Lankan users: A study on password management and adoption of password Managers. *International Journal of Education and Management Engineering*, *14*(1), 33–40. DOI: 10.5815/ijeme.2024.01.04

Fredrikson, M., Jha, S., & Ristenpart, T. (2015). Model inversion attacks that exploit confidence information and basic countermeasures. In *Proceedings of the 22nd ACM SIGSAC Conference on Computer and Communications Security* (pp. 1322–1333). DOI: 10.1145/2810103.2813677

Fujak, H., Karg, A., & Whitburn, D. (2022). Sport media has an audience measurement problem. *European Sport Management Quarterly*, *24*(3), 582–602. DOI: 10.1080/16184742.2022.2153376

Gallo, A. J., Turan, M. S., Boem, F., Parisini, T., & Ferrari-Trecate, G. (2020). A distributed cyber-attack detection scheme with application to DC microgrids. *IEEE Transactions on Automatic Control*, *65*(9), 3800–3815. DOI: 10.1109/TAC.2020.2982577

Gao, F., Yu, C., Zhou, H., & Xie, L. (2024). Behavioral lock screen application practice based on Gamification Design method. In *Lecture notes in computer science* (pp. 43–57). DOI: 10.1007/978-3-031-60898-8_3

Garrett, G. A. (2018). *Cybersecurity in the digital age: Tools, Techniques, & Best Practices*. Aspen Publishers.

Gasiewski, M. (2024). Revolutionizing revenue at Google. In *Routledge eBooks* (pp. 43–49). DOI: 10.4324/9781003535966-5

Ghelani, D., Hua, T. K., & Koduru, S. K. R. (2022). A Model-Driven Approach for Online Banking Application Using AngularJS Framework. *American Journal of Information Science and Technology*, *6*(3), 52–63.

Ghorpade, S. N., Zennaro, M., & Chaudhari, B. S. (2021). IoT-based hybrid optimized fuzzy threshold ELM model for localization of elderly persons. *Expert Systems with Applications*, *184*, 115500. DOI: 10.1016/j.eswa.2021.115500

Gil-Aciron, L. A. (2022). The gamer psychology: A psychological perspective on game design and gamification. *Interactive Learning Environments*, *32*(1), 183–207. DOI: 10.1080/10494820.2022.2082489

Gill, S., & Devi, R. (2024). Enhancing Cloud Data Security using Artificial Neural Networks for Users' Account Hijacking Security Threats. *Indian Journal of Science and Technology*, *17*(34), 3538–3552. DOI: 10.17485/IJST/v17i34.2339

Gnanasekaran, S. M. (2024). *User privacy in OAUth-Based Single Sign-On systems*. DOI: 10.22215/etd/2024-15946

Gong, M., Wagner, C., & Ali, A. (2023). The impact of social network embeddedness on mobile massively multiplayer online games play. *Information Systems Journal*, *34*(2), 327–363. DOI: 10.1111/isj.12479

Gopi, R., Sathiyamoorthi, V., Selvakumar, S., Manikandan, R., Chatterjee, P., Jhanjhi, N. Z., & Luhach, A. K. (2021). Enhanced method of ANN based model for detection of DDoS attacks on multimedia internet of things. *Multimedia Tools and Applications*, *81*(19), 26739–26757. DOI: 10.1007/s11042-021-10640-6

Gouda, W., Almurafeh, M., Humayun, M., & Jhanjhi, N. Z. (2022). Detection of COVID-19 based on chest x-rays using deep learning. *Health Care*, *10*(2), 343. DOI: 10.3390/healthcare10020343 PMID: 35206957

Gouveia, A., & Correia, M. (2020). Towards quantum-enhanced machine learning for network intrusion detection. In *2020 IEEE 19th International Symposium on Network Computing and Applications* (pp. 1–8). DOI: 10.1109/NCA51143.2020.9306691

Griebel, L., Prokosch, H.-U., Köpcke, F., Toddenroth, D., Christoph, J., Leb, I., Engel, I., & Sedlmayr, M. (2015). A scoping review of cloud computing in healthcare. *BMC Medical Informatics and Decision Making*, *15*(1), 17. Advance online publication. DOI: 10.1186/s12911-015-0145-7 PMID: 25888747

Guido, M., Ondricek, J., Grover, J., Wilburn, D., Nguyen, T., & Hunt, A. (2013). Automated identification of installed malicious Android applications. *Digital Investigation*, *10*, S96–S104. DOI: 10.1016/j.diin.2013.06.011

Gui, J., Mcilroy, S., Nagappan, M., & Halfond, W. G. J. (2015). Truth in advertising: the hidden cost of mobile ads for software developers. *International Conference on Software Engineering*, *1*, 100–110. DOI: 10.1109/ICSE.2015.32

Guluzade, L., & Sas, C. (2024). Functionality and user review analysis of mobile apps for mindfulness eating and eating disorders. *Designing Interactive Systems Conference*, *6*, 1350–1371. DOI: 10.1145/3643834.3661521

Gupta, S., Maple, C., Crispo, B., Raja, K., Yautsiukhin, A., & Martinelli, F. (2023a). A survey of human-computer interaction (HCI) & natural habits-based behavioural biometric modalities for user recognition schemes. *Pattern Recognition*, *139*, 109453. DOI: 10.1016/j.patcog.2023.109453

Güven, E. Y., Boyaci, A., & Aydin, M. A. (2021). A novel password policy focusing on altering user password selection habits: A statistical analysis on breached data. *Computers & Security, 113*, 102560. DOI: 10.1016/j.cose.2021.102560

Hagui, I., Msolli, A., Henda, N. B., Helali, A., Gassoumi, A., Nguyen, T. P., & Hassen, F. (2023). A blockchain-based security system with light cryptography for user authentication security. *Multimedia Tools and Applications, 83*(17), 52451–52480. DOI: 10.1007/s11042-023-17643-5

Hall-Johnson, T., & Hall-Johnson, L. (2023). *Business model : FocusHome*. Theseus. https://www.theseus.fi/handle/10024/801120

Hammontree, M., Weiler, P., & Nayak, N. (1994). Remote usability testing. *interactions, 1*(3), 21-25.

Hanif, Y., & Lallie, H. S. (2021). Security factors on the intention to use mobile banking applications in the UK older generation (55+). A mixed-method study using modified UTAUT and MTAM-with perceived cyber security, risk, and trust. *Technology in Society, 67*, 101693. DOI: 10.1016/j.techsoc.2021.101693

Harbinger Group. (2024, April 18). *How to build a gamification Strategy – An expert view*. https://www.harbingergroup.com/blogs/how-to-build-a-gamification-strategy-an-expert-view/

Harisha, N., Naik, S. R., Vasudeva, S. S., Shrilakshmi, K., & Kothwal, V. (2024). Advancements in User Security: Enhancing Usability with Graphical Password Authentication. *2024 2nd International Conference on Intelligent Data Communication Technologies and Internet of Things (IDCIoT)*. DOI: 10.1109/IDCIoT59759.2024.10467993

Harmening, J. (2025). Security Management Systems. In *Elsevier eBooks* (pp. 433–439). DOI: 10.1016/B978-0-443-13223-0.00025-4

He, D., Liu, Z., Zhu, S., Chan, S., & Guizani, M. (2024). Special characters usage and its effect on password security. *IEEE Internet of Things Journal, 11*(11), 19440–19453. DOI: 10.1109/JIOT.2024.3367323

Hellín, C. J., Calles-Esteban, F., Valledor, A., Gómez, J., Otón-Tortosa, S., & Tayebi, A. (2023). Enhancing Student Motivation and Engagement through a Gamified Learning Environment. *Sustainability (Basel), 15*(19), 14119. DOI: 10.3390/su151914119

Hennebelle, A., Materwala, H., & Ismail, L. (2023). HealthEdge: A machine learning-based smart healthcare framework for prediction of type 2 diabetes in an integrated IoT, edge, and cloud computing system. *Procedia Computer Science, 220*, 331–338. DOI: 10.1016/j.procs.2023.03.043

Herbach, C. (2024). Placing people at the heart of customer experience journeys. In *Routledge eBooks* (pp. 65–92). DOI: 10.4324/9781003378099-5

He, Z., Huq, S. F., & Malek, S. (2024). *"I tend to view ads almost like a pestilence": On the Accessibility Implications of Mobile Ads for Blind Users*. ACM., DOI: 10.1145/3597503.3639228

Hochstein, B., Voorhees, C. M., Pratt, A. B., Rangarajan, D., Nagel, D. M., & Mehrotra, V. (2023). Customer success management, customer health, and retention in B2B industries. *International Journal of Research in Marketing*, *40*(4), 912–932. DOI: 10.1016/j.ijresmar.2023.09.002

Hong, Y., Saab, N., & Admiraal, W. (2024). Approaches and game elements used to tailor digital gamification for learning: A systematic literature review. *Computers & Education*, *212*, 105000. DOI: 10.1016/j.compedu.2024.105000

Hsu, S., & Karahalios, K. (2024). *Choosing What You Want Versus Getting What You Want: An Experiment with Choice in Video Ad Placement*. ACM., DOI: 10.1145/3613904.3642869

Huang, J., Chen, J., & Zhou, L. (2024). Motivation crowding effects on the intention for continued use of gamified fitness apps: A mixed-methods approach. *Frontiers in Psychology*, *14*, 1286463. Advance online publication. DOI: 10.3389/fpsyg.2023.1286463 PMID: 38268805

Humayun, M., Sujatha, R., Almuayqil, S. N., & Jhanjhi, N. Z. (2022). A Transfer Learning Approach with a Convolutional Neural Network for the Classification of Lung Carcinoma. *Health Care*, *10*(6), 1058. DOI: 10.3390/healthcare10061058 PMID: 35742109

Hunt, A., Merola, G. P., Carpenter, T., & Jaeggi, A. V. (2024). Evolutionary perspectives on Substance and Behavioural Addictions: Distinct and shared pathways to understanding, prediction and prevention. *Neuroscience and Biobehavioral Reviews*, *105603*, 105603. Advance online publication. DOI: 10.1016/j.neubiorev.2024.105603 PMID: 38402919

Hussain, K., Rahmatyar, A. R., Riskhan, B., Sheikh, M. U., & Sindiramutty, S. R. (2024). Threats and Vulnerabilities of Wireless Networks in the Internet of Things (IoT). *2024 IEEE 1st Karachi Section Humanitarian Technology Conference (KHI-HTC), 2*, 1–8. DOI: 10.1109/KHI-HTC60760.2024.10482197

Hussain, A., Ting, D. H., & Marder, B. (2024). Why premium in freemium: A hedonic shopping motivation model in virtual game retailing. *Information Technology & People*. Advance online publication. DOI: 10.1108/ITP-01-2023-0082

Hwaitat, A. K. A., Fakhouri, H. N., Alawida, M., Atoum, M. S., Abu-Salih, B., Salah, I. K. M., Al-Sharaeh, S., & Alassaf, N. (2024). Overview of mobile attack detection and prevention techniques using Machine Learning. [iJIM]. *International Journal of Interactive Mobile Technologies, 18*(10), 125–157. DOI: 10.3991/ijim.v18i10.46485

Ibisu, A. E. (2024). Development of a gamification model for personalized e-learning. *arXiv (Cornell University)*. https://doi.org//arxiv.2404.15301DOI: 10.48550

Igwenagu, U. T. I., Salami, A. A., Arigbabu, A. S., Mesode, C. E., Oladoyinbo, T. O., & Olaniyi, O. O. (2024). Securing the digital frontier: Strategies for cloud computing security, database protection, and comprehensive penetration testing. *Journal of Engineering Research and Reports, 26*(6), 60–75. DOI: 10.9734/jerr/2024/v26i61162

Ijomah, N. T. I., Idemudia, N. C., Eyo-Udo, N. N. L., & Anjorin, N. K. F. (2024). The role of big data analytics in customer relationship management: Strategies for improving customer engagement and retention. *World Journal of Advanced Science and Technology, 6*(1), 013–024. DOI: 10.53346/wjast.2024.6.1.0038

Isha, L., A. K., & Kumar, S. (2016, August). Layer based security in Internet of Things: current mechanisms, prospective attacks, and future orientation. In International Conference on Smart Trends for Information Technology and Computer Communications (pp. 896-903). Singapore: Springer Nature Singapore.

Islam, U., Muhammad, A., Mansoor, R., Hossain, M. S., Ahmad, I., Eldin, E. T., Khan, J. A., Rehman, A. U., & Shafiq, M. (2022). Detection of distributed denial of service (DDoS) attacks in IOT based monitoring system of banking sector using machine learning models. *Sustainability (Basel), 14*(14), 8374. DOI: 10.3390/su14148374

Jain, C., Sashank, G. V., Venkateswaran, N., & Markkandan, S. (2021). Low-cost BLE based indoor localization using RSSI fingerprinting and machine learning. In *Proceedings of the 2021 Sixth International Conference on Wireless Communications, Signal Processing and Networking (WiSPNET)* (pp. 363–367). Chennai, India. DOI: 10.1109/WiSPNET51692.2021.9419388

Jamwal, S., Cano, J., Lee, G. M., Tran, N. H., & Truong, N. (2024). A survey on ethereum pseudonymity: Techniques, challenges, and future directions. *Journal of Network and Computer Applications, 232*, 104019. DOI: 10.1016/j.jnca.2024.104019

Jansukpum, K., Chernbumroong, S., Intawong, K., Sureephong, P., & Puritat, K. (2024). Gamified Virtual Reality for Library services: The effect of gamification on enhancing knowledge retention and user engagement. *New Review of Academic Librarianship*, •••, 1–21. DOI: 10.1080/13614533.2024.2381509

Javaid, S., Ashraf, H., & Jhanjhi, N. (2023, December 21). Efficient cryptographic technique for data protection of wireless body area network. DOI: 10.20944/preprints202312.1609.v1

Jelinski, J. (2024). *Needle work: A History of Commercial Tattooing in Canada Volume 44*.

Jha, A. K., & Verma, N. K. (2023). Social Media Platforms and User Engagement: A Multi-Platform Study on one-way firm sustainability communication. *Information Systems Frontiers*, *26*(1), 177–194. DOI: 10.1007/s10796-023-10376-8

Jhanjhi, N., Humayun, M., & Almuayqil, S. N. (2021). Cyber security and privacy issues in industrial internet of things. *Computer Systems Science and Engineering*, *37*(3), 361–380. DOI: 10.32604/csse.2021.015206

Jia, B., Zong, Z., Huang, B., & Baker, T. (2021). A DNN-based WiFi-RSSI indoor localization method in IoT. In *Springer Science and Business Media Deutschland GmbH* (Vol. 352, pp. 200–211). Berlin/Heidelberg, Germany. DOI: 10.1007/978-3-030-67720-6_14

Jiang, X., Dou, R., He, Q., Zhang, X., & Dou, W. (2023). EdgeAuth: An intelligent token-based collaborative authentication scheme. *Software, Practice & Experience*, *54*(10), 2007–2024. DOI: 10.1002/spe.3206

Josyula, H. P. (2023). Fraud detection in fintech leveraging machine learning and behavioral analytics. *Research Square (Research Square)*. DOI: 10.21203/rs.3.rs-3548343/v1

Juhász, L., Novack, T., Hochmair, H., & Qiao, S. (2020). Cartographic vandalism in the era of Location-Based Games—The case of OpenStreetMap and Pokémon GO. *ISPRS International Journal of Geo-Information*, *9*(4), 197. DOI: 10.3390/ijgi9040197

Jun, A. Y. M., Jinu, B. A., Seng, L. K., Maharaiq, M. H. F. B. Z., Khongsuwan, W., Junn, B. T. K., Hao, A. W., & Sindiramutty, S. R. (2024). Exploring the Impact of Crypto-Ransomware on Critical Industries: Case Studies and Solutions. *Preprints.org*. DOI: 10.20944/preprints202409.1325.v1

Kammoun, A., Slama, R., Tabia, H., Ouni, T., & Abid, M. (2022a). Generative Adversarial Networks for face generation: A survey. *ACM Computing Surveys*, 1122445.1122456. Advance online publication. DOI: 10.1145/1122445.1122456

Kanungo, K., Khatoliya, R., Arora, V., Bari, A., Bhattacharya, A., Maity, M., & Chakravarty, S. (2024). How Many Hands in the Cookie Jar? Examining Privacy Implications of Popular Apps in India. *2024 IEEE 9th European Symposium on Security and Privacy (EuroS&P)*, 741–757. DOI: 10.1109/EuroSP60621.2024.00046

Karim, N. A., Kanaker, H., Abdulraheem, W. K., Ghaith, M. A., Alhroob, E., & Alali, A. M. F. (2024). Choosing the right MFA method for online systems: A comparative analysis. *International Journal of Data and Network Science*, *8*(1), 201–212. DOI: 10.5267/j.ijdns.2023.10.003

Karpouzis, K. (2024). What would Plato say? Concepts and notions from Greek philosophy applied to gamification mechanics for a meaningful and ethical gamification. *arXiv (Cornell University)*. /arxiv.2403.08041DOI: 10.31219/osf.io/u4cqn

Kastelli, S., & Takács, N. (2023). *Persuasive Language Learning - Qualitative research on user engagement in the persuasive system design of Duolingo*. DIVA. https://www.diva-portal.org/smash/record.jsf?pid=diva2:1773780

Katariya, J. (2024, June 19). 15 Mobile App Monetization Strategies for Maximizing Revenue. *Moon Technolabs*. https://www.moontechnolabs.com/blog/mobile-app-monetization/

Kaur, A. A., & Mustafa, K. K. (2019). A Critical appraisal on Password based Authentication. *International Journal of Computer Network and Information Security*, *11*(1), 47–61. DOI: 10.5815/ijcnis.2019.01.05

Kaushik, K., Ouaissa, M., & Chaudhary, A. (2024). *Advanced techniques and applications of cybersecurity and forensics*. CRC Press. DOI: 10.1201/9781003386926

Kavitha, G., Prasannakumar, V., & Pranav, R. (2024). Enhancing Digital Security: A Comprehensive Multi-Model Authentication Framework Leveraging Cryptography and Biometrics. *2024 8th International Conference on Inventive Systems and Control (ICISC)*, 476–486. DOI: 10.1109/ICISC62624.2024.00086

Kävrestad, J., Birath, M., & Clarke, N. (2024). Decryption and password cracking. In *Texts in computer science* (pp. 43–53). DOI: 10.1007/978-3-031-53649-6_5

Khamaj, A., & Ali, A. M. (2024). Adapting user experience with reinforcement learning: Personalizing interfaces based on user behavior analysis in real-time. *Alexandria Engineering Journal*, *95*, 164–173. DOI: 10.1016/j.aej.2024.03.045

Khamisa, A. (2021). Token economies. In *Emerald Publishing Limited eBooks* (pp. 121–140). DOI: 10.1108/978-1-83982-198-120211011

Khan, J. A. (2024). Role-Based access Control (RBAC) and Attribute-Based Access Control (ABAC). In *Advances in information security, privacy, and ethics book series* (pp. 113–126). DOI: 10.4018/979-8-3693-1431-9.ch005

Khan, A., Jhanjhi, N. Z., & Humayun, M. (2022). The role of cybersecurity in smart cities. In *Cyber Security Applications for Industry 4.0* (pp. 195–208). Chapman and Hall/CRC. DOI: 10.1201/9781003203087-9

Khan, A., Jhanjhi, N., Abdulhabeb, G. A. A., Ray, S. K., Ghazanfar, M. A., & Humayun, M. (2025). Securing IoT Devices Using Generative AI Techniques. In *Reshaping CyberSecurity With Generative AI Techniques* (pp. 219–264). IGI Global.

Khan, H. U., Sohail, M., Nazir, S., Hussain, T., Shah, B., & Ali, F. (2023). Role of authentication factors in Fin-tech mobile transaction security. *Journal of Big Data*, *10*(1), 138. DOI: 10.1186/s40537-023-00807-3

Khan, S. (2024). *Brave new words: How AI Will Revolutionize Education (and Why That's a Good Thing)*. Random House.

Khan, S., Devlen, C., Manno, M., & Hou, D. (2024). Mouse Dynamics Behavioral Biometrics: A survey. *ACM Computing Surveys*, *56*(6), 1–33. DOI: 10.1145/3640311

Kim, S., Nguyen, L. T., Kim, J., & Shim, B. (2021). Deep learning based low-rank matrix completion for IoT network localization. *IEEE Wireless Communications Letters*, *10*(10), 2115–2119. DOI: 10.1109/LWC.2021.3093983

Kircher, T., & Foerderer, J. (2023). Ban Targeted Advertising? An Empirical investigation of the consequences for app development. *Management Science*, *70*(2), 1070–1092. DOI: 10.1287/mnsc.2023.4726

Kissell, J. (2024). Take Control of Your Passwords, *4th Edition*. alt concepts.

Kiyani, F. F., Hamid, B., Humayun, M., Sindiramutty, S. R. L., & Chowdhury, S. (2024). Discovery of Influential Publications Using Research Article's Usage Context. *2024 International Conference on Emerging Trends in Networks and Computer Communications (ETNCC)*, 1–7. DOI: 10.1109/ETNCC63262.2024.10767576

Kizza, J. M. (2024a). Access control and authorization. In *Texts in computer science* (pp. 195–214). DOI: 10.1007/978-3-031-47549-8_9

Kizza, J. M. (2024b). Authentication. In *Texts in computer science* (pp. 215–238). DOI: 10.1007/978-3-031-47549-8_10

Kliegr, T., & Izquierdo, E. (2023). QCBA: Improving rule classifiers learned from quantitative data by recovering information lost by discretisation. *Applied Intelligence*, *53*(18), 20797–20827. DOI: 10.1007/s10489-022-04370-x

Kok, S. H., Abdullah, A., & Jhanjhi, N. (2022, May). Early detection of crypto-ransomware using pre-encryption detection algorithm. *Journal of King Saud University. Computer and Information Sciences*, *34*(5), 1984–1999. DOI: 10.1016/j.jksuci.2020.06.012

Kong, A., & Feng, Z. (2024). Advancing VR edutainment design in blended learning: Learners' views from wine classroom. *Computers & Education X Reality, 5*, 100078. DOI: 10.1016/j.cexr.2024.100078

Koolen, C., Wuyts, K., Joosen, W., & Valcke, P. (2024). From insight to compliance: Appropriate technical and organisational security measures through the lens of cybersecurity maturity models. *Computer Law & Security Review*, *52*, 105914. DOI: 10.1016/j.clsr.2023.105914

Kordi, K. A., Alhammadi, A., Roslee, M., & Alias, M. Y. (2020). A review on wireless emerging IoT indoor localization. In *Proceedings of the 2020 IEEE 5th International Symposium on Telecommunication Technologies (ISTT)* (pp. 82–87). Shah Alam, Malaysia. DOI: 10.1109/ISTT50966.2020.9279386

Kowalkowski, C., & Ulaga, W. (2024). Subscription offers in business-to-business markets: Conceptualization, taxonomy, and framework for growth. *Industrial Marketing Management, 117*, 440–456. DOI: 10.1016/j.indmarman.2024.01.014

Ksiezopolski, B., Mazur, K., Miskiewicz, M., & Rusinek, D. (2022). Teaching a Hands-On CTF-Based Web Application Security course. *Electronics (Basel)*, *11*(21), 3517. DOI: 10.3390/electronics11213517

Kumar, M. S., Vimal, S., Jhanjhi, N., Dhanabalan, S. S., & Alhumyani, H. A. (2021). Blockchain based peer to peer communication in autonomous drone operation. *Energy Reports*, *7*, 7925–7939. DOI: 10.1016/j.egyr.2021.08.073

Kumar, V., Malik, N., Singla, J., Jhanjhi, N. Z., Amsaad, F., & Razaque, A. (2022, July). Light Weight Authentication Scheme for Smart Home IoT Devices. *Cryptography*, *6*(3), 37. DOI: 10.3390/cryptography6030037

Kun, E. (2024). Challenges in regulating cloud service providers in EU financial regulation: From operational to systemic risks, and examining challenges of the new oversight regime for critical cloud service providers under the Digital Operational Resilience Act. *Computer Law & Security Review*, *52*, 105931. DOI: 10.1016/j.clsr.2023.105931

Kunkel, T., Hayduk, T., & Lock, D. (2023). Push it real good: The effects of push notifications promoting motivational affordances on consumer behavior in a gamified mobile app. *European Journal of Marketing*, *57*(9), 2592–2618. DOI: 10.1108/EJM-06-2021-0388

Lacko, D., Machackova, H., & Slavík, L. (2024). Adolescents' perceptions of the credibility of informational content on fitness and dietary supplements: The impact of banner and native advertising. *Journal of Adolescence*, *96*(8), 1956–1968. Advance online publication. DOI: 10.1002/jad.12394 PMID: 39164994

Laine, T. H. (2018). Mobile Educational Augmented Reality Games: A systematic literature review and two case studies. *Computers*, *7*(1), 19. DOI: 10.3390/computers7010019

Lakshmi, K. K., Gupta, H., & Ranjan, J. (2019). UPI based Mobile Banking Applications – Security Analysis and enhancements. *2019 Amity International Conference on Artificial Intelligence (AICAI)*, 1–6. DOI: 10.1109/AICAI.2019.8701396

Landers, R. N., Bauer, K. N., Callan, R. C., & Armstrong, M. B. (2014). Psychological theory and the gamification of learning. In *Springer eBooks* (pp. 165–186). DOI: 10.1007/978-3-319-10208-5_9

Lange, P. G. (2024). Ranting in emotional public spheres: Publicizing participatory challenges on YouTube. *Convergence (London)*, *30*(2), 904–921. DOI: 10.1177/13548565241246044

Larsen, N., Stallrich, J., Sengupta, S., Deng, A., Kohavi, R., & Stevens, N. T. (2023). Statistical challenges in online controlled experiments: A review of A/B testing methodology. *The American Statistician*, •••, 1–15. DOI: 10.1080/00031305.2023.2257237

Lasén, A. (2004). *Affective technologies—emotions and mobile phones*. The Digital World Research Centre.

Lee, M. (2024). *Hacks, leaks, and revelations: The Art of Analyzing Hacked and Leaked Data*. No Starch Press.

Lee, S., Yang, Z., & Wu, J. G. (2023). Live, play, and learn: Language learner engagement in the immersive VR environment. *Education and Information Technologies*, *29*(9), 10529–10550. DOI: 10.1007/s10639-023-12215-4

Lemos, N., Machado, C. S., & Cardoso, C. (2023). Free apps and paid apps: Monetization strategies for health apps in the Portuguese market. *International Journal of Pharmaceutical and Healthcare Marketing*, *18*(2), 237–251. DOI: 10.1108/IJPHM-01-2023-0001

Liang, S., Wei, Z., & Zang, L. (2024). An exploratory study of factors influencing user app abandonment on smartphones. *Library Hi Tech*. Advance online publication. DOI: 10.1108/LHT-07-2022-0349

Li, J., Cochrane, K. A., & Leshed, G. (2024). Beyond meditation: understanding everyday mindfulness practices and technology use among experienced practitioners. *Proceedings of the ACM on Human-Computer Interaction, 8*(CSCW2), 1–29. DOI: 10.1145/3687023

Li, J., Wu, J., Zhang, J., & Ji, X. (2024). Freemium design: Optimal tier differentiation models for content platforms. *Transportation Research Part E, Logistics and Transportation Review*, *188*, 103612. DOI: 10.1016/j.tre.2024.103612

Li, L., Lu, X., & Wang, K. (2022, December). Hash-based signature revisited. *Cybersecurity*, *5*(1), 13. DOI: 10.1186/s42400-022-00117-w

Lim, M., Abdullah, A., Jhanjhi, N., Khan, M. K., & Supramaniam, M. (2019). Link Prediction in Time-Evolving Criminal Network with deep Reinforcement learning technique. *IEEE Access : Practical Innovations, Open Solutions*, *7*, 184797–184807. DOI: 10.1109/ACCESS.2019.2958873

Lindgren, E. (2024). *Accessibility redesign and its effect on purchase intent and customer experience*. DIVA. https://www.diva-portal.org/smash/record.jsf?pid=diva2%3A1864084&dswid=-285

Lindström, C. W. J., Vishkaei, B. M., & De Giovanni, P. (2023). Subscription-based business models in the context of tech firms: theory and applications. *International Journal of Industrial Engineering and Operations Management*. DOI: 10.1108/IJIEOM-06-2023-0054

Linqiang, Y., Sindiramutty, S. R. L., Ashraf, H., Muzammal, S. M., Balakrishnan, S. P., Gupta, S., & Kavita, N. (2024). Intelligent Household Waste Classification System Based on Machine Learning. *2024 International Conference on Emerging Trends in Networks and Computer Communications (ETNCC)*, 760–768. DOI: 10.1109/ETNCC63262.2024.10767563

Liu, Y., He, J., Li, X., Chen, J., Liu, X., Peng, S., Cao, H., & Wang, Y. (2024). An overview of blockchain smart contract execution mechanism. *Journal of Industrial Information Integration*, *41*, 100674. DOI: 10.1016/j.jii.2024.100674

Li, X., Balasubramanian, H., Chen, Y., & Pang, C. (2023). Managing conflicting revenue streams from advertisers and subscribers for online platforms. *European Journal of Operational Research*, *314*(1), 241–254. DOI: 10.1016/j.ejor.2023.10.024

Li, X., Yang, Y., & Chu, S. K. W. (2024). How does gamification bring long-term sustainable effects on children's learning? Implications from a crossover quasi-experimental study. *Educational Technology Research and Development*, *72*(3), 1357–1381. Advance online publication. DOI: 10.1007/s11423-023-10341-x

Li, Y.. (2020). Toward location-enabled IoT (LE-IoT): IoT positioning techniques, error sources, and error mitigation. *IEEE Internet of Things Journal*, *8*(5), 4035–4062. DOI: 10.1109/JIOT.2020.3021611

Lu, J., Bradlow, E. T., & Hutchinson, J. W. (2024). More Likely to Pay but Less Engaged: The Effects of Switching Online Courses from Scheduled to On-Demand Release on User Behavior. *Journal of Marketing*, *88*(5), 63–88. DOI: 10.1177/00222429241227145

Lut, L. T. Y. (2024). *Utilizing customer preferences to enhance digital services : appealing features of subscription model and subscription management portal*. LUTPub. https://lutpub.lut.fi/handle/10024/168439

Lüthy, K. (2024). Periodicals/Subject groups. In *Springer eBooks* (pp. 765–789). DOI: 10.1007/978-3-658-39909-2_33

Lykousas, N., & Patsakis, C. (2024). Decoding developer password patterns: A comparative analysis of password extraction and selection practices. *Computers & Security*, *145*, 103974. DOI: 10.1016/j.cose.2024.103974

Maglogiannis, I. (2024). *Artificial Intelligence Applications and Innovations. AIAI 2024 IFIP WG 12.5 International Workshops: MHDW 2024, 5G-PINE 2024, and AI4GD 2024, Corfu, Greece, June 27–30, 2024, Proceedings*. Springer Nature.

Mahmood, M. R., Matin, M. A., Sarigiannidis, P., & Goudos, S. K. (2022). A comprehensive review on artificial intelligence/machine learning algorithms for empowering the future IoT toward 6G era. *IEEE Access : Practical Innovations, Open Solutions*, *10*, 87535–87562. DOI: 10.1109/ACCESS.2022.3199689

Mahmoudi-Dehaki, M., & Nasr-Esfahani, N. (2024). Harnessing the power of persuasion in education. In *Advances in educational technologies and instructional design book series* (pp. 177–204). DOI: 10.4018/979-8-3693-6397-3.ch008

Malhotra, Y. (2022, June). *How you can implement well-architected 'zero trust' hybrid-cloud computing beyond 'lift and shift': Cloud-enabled digital innovation at scale with infrastructure as code (IaC), DevSecOps, and MLOps. In 2022 New York State Cyber Security Conference*. Invited Presentations., https://its.ny.gov/2022-nyscsc

Malik, G., Pradhan, D., & Rup, B. K. (2024). Gamification and customer brand engagement: A review and future research agendas. *Marketing Intelligence & Planning*. Advance online publication. DOI: 10.1108/MIP-12-2023-0659

Mali, S. (2024). Assessing the effectiveness of Multi-Factor authentication in Cloud-Based big data environments. *Internet of Things and Cloud Computing, 12*(2), 17–27. DOI: 10.11648/j.iotcc.20241202.11

Malki, L. M., Kaleva, I., Patel, D., Warner, M., & Abu-Salma, R. (2024). *Exploring Privacy Practices of Female mHealth Apps in a Post-Roe World.* ACM., DOI: 10.1145/3613904.3642521

Manchuri, A., Kakera, A., Saleh, A., & Raja, S. (2024). pplication of Supervised Machine Learning Models in Biodiesel Production Research - A Short Review. *Borneo Journal of Sciences and Technology*. DOI: 10.35370/bjost.2024.6.1-10

Mazlan, A. M. (2024). Savings vs Splurges : Mindful consumption for financial well-being. SSRN *Electronic Journal*. DOI: 10.2139/ssrn.4844943

McKay, E. N. (2013). *UI is Communication: How to design intuitive, user centered interfaces by focusing on effective communication*. https://dl.acm.org/citation.cfm?id=2513625

McKinnon, M. M. (2023). *Ad or Editorial? Native Advertising and Shifting Political Attitudes - ProQuest.* https://www.proquest.com/openview/1a3e962f1720be3b291866b43ff9be2d/1?cbl=18750&diss=y&pq-origsite=gscholar

Mekler, E. D., Brühlmann, F., Opwis, K., & Tuch, A. N. (2013). *Do points, levels and leaderboards harm intrinsic motivation?* ACM., DOI: 10.1145/2583008.2583017

Mellado, R. (2024). *Principles of gamification for educational software*. Editora Dialética.

Melzi, P., Rathgeb, C., Tolosana, R., Vera, R., & Busch, C. (2024). An Overview of Privacy-Enhancing Technologies in biometric Recognition. *ACM Computing Surveys, 56*(12), 1–28. Advance online publication. DOI: 10.1145/3664596

Mihale-Wilson, C., & Carl, K. V. (2024). Designing incentive systems for participation in digital ecosystems—An integrated framework. *Electronic Markets, 34*(1), 16. Advance online publication. DOI: 10.1007/s12525-024-00703-5

Mirtaheri, S. L., & Shahbazian, R. (2022). *Machine learning: Theory to applications*. CRC Press. DOI: 10.1201/9781003119258

Modarelli, G., & Rainero, C. (2024). Perceptions, Criticalities, and Potentials of Game-Based Trainings for Inclusion. In *Advances in business information systems and analytics book series* (pp. 1–48). DOI: 10.4018/979-8-3693-1172-1.ch001

Mohammed, N., Saleem, A. N. S., & Anwar, M. (2024). Gamification application for promoting and encouraging physical activity in the elderly. eajse.tiu.edu.iq. DOI: 10.23918/eajse.v10i2p16

Mohanty, S., Tripathy, A., & Das, B. (2021). An overview of a low energy UWB localization in IoT based system. In *Proceedings of the 2021 International Symposium of Asian Control Association on Intelligent Robotics and Industrial Automation (IRIA)* (pp. 293–296). Goa, India. DOI: 10.1109/IRIA53009.2021.9588718

Moller, A. C., Kornfield, R., & Lu, A. S. (2024). Competition and Digital Game Design: A Self-Determination Theory Perspective. *Interacting with Computers*, iwae023. Advance online publication. DOI: 10.1093/iwc/iwae023

Monteiro, M. M., De Azevedo, C. M. L., Kamargianni, M., Cantelmo, G., Tavory, S. S., Gal-Tzur, A., Antoniou, C., & Shiftan, Y. (2023). Car-sharing subscription preferences and the role of incentives: The case of Copenhagen, Munich, and Tel Aviv-Yafo. *Case Studies on Transport Policy*, *12*, 101013. DOI: 10.1016/j.cstp.2023.101013

Moradbeiki, A., Keshavarz, A., Rostami, H., Paiva, S., & Lopes, S. I. (2021). GNSS-free outdoor localization techniques for resource-constrained IoT architectures: A literature review. *Applied Sciences (Basel, Switzerland)*, *11*(23), 10793. DOI: 10.3390/app112210793

Moradzadeh, S., & Kou, Y. (2024). "Wow another fake game from YouTube ad": Unpacking Fake Games Through a Mixed-Methods Investigation. *Proceedings of the ACM on Human-Computer Interaction*, *8*(CHI PLAY), 1–36. DOI: 10.1145/3677115

Morais, R. A. (2024). *Design & Advertising: Applications of design in advertising departments*. Editora Appris.

Morales, M. E. S., Tlyachev, T., & Biamonte, J. (2018, December). Variational learning of Grover's quantum search algorithm. *Physical Review. A*, *98*(6), 062333. DOI: 10.1103/PhysRevA.98.062333

Morisio, M. (n.d.). *Design and implementation of a comfortable Mobile-based monitoring and recommendation system for healthy lifestyle habits - Webthesis*. https://webthesis.biblio.polito.it/31855/

Mutlag, A. A., Abd Ghani, M. K., Mohammed, M. A., Lakhan, A., Mohd, O., Abdulkareem, K. H., & Garcia-Zapirain, B. (2021). Multi-agent systems in fog–cloud computing for critical healthcare task management model (CHTM) used for ECG monitoring. *Sensors (Basel), 21*(20), 6923. DOI: 10.3390/s21206923 PMID: 34696135

Nabati, M., & Ghorashi, S. A. (2020). Joint coordinate optimization in fingerprint-based indoor positioning. *IEEE Communications Letters*, *25*(5), 1192–1195. DOI: 10.1109/LCOMM.2020.3042375

Najat, R., Haitam, E., & Jaafar, A. (2023). Comparative study of the Security Analysis of IoT systems using attack trees algorithm. *E3S Web of Conferences, 412*, 01087. DOI: 10.1051/e3sconf/202341201087

Nanda, A., Jeong, J. J., Shah, S. W. A., Nosouhi, M., & Doss, R. (2023). Examining usable security features and user perceptions of physical authentication devices. *Computers & Security, 139*, 103664. DOI: 10.1016/j.cose.2023.103664

Nasser, N., Emad-ul-Haq, Q., Imran, M., Ali, A., Razzak, I., & Al-Helali, A. (2021). A smart healthcare framework for detection and monitoring of COVID-19 using IoT and cloud computing. *Neural Computing & Applications*, •••, 1–15. PMID: 34522068

Natalie, R., Chang, R., Sheshadri, S., Guo, A., & Hara, K. (2024)... *Audio Description Customization. ACM, 84*, 1–19. DOI: 10.1145/3663548.3675617

Nayyar, A., Gadhavi, L., & Zaman, N. (2021). Machine learning in healthcare: review, opportunities and challenges. In *Elsevier eBooks* (pp. 23–45). DOI: 10.1016/B978-0-12-821229-5.00011-2

Neuschafer, T. (2023). *Understanding Duolingo Discussion Boards as Social-Emotional Support during the COVID-19 Pandemic.* https://eric.ed.gov/?id=EJ1393729

Nigam, A. C., & Chanda, R. S. (2024). Digital Innovations and their Effect on user Interaction in fitness Apps: A Comprehensive review and future research directions. *Advanced Series in Management*, 113–135. DOI: 10.1108/S1877-636120240000034B010

Noble, B. D., & Corner, M. D. (2002). *The case for transient authentication.* ACM., DOI: 10.1145/1133373.1133377

Norizan, A. R., Kamil, M. H. F. M., Zamri, M. F. M., & Khalid, F. S. (2024). Comparing In-Game purchasing behavior between youth and adult gamers: Understanding enjoyment rates and implications for game developers. In *Studies in systems, decision and control* (pp. 303–316). DOI: 10.1007/978-3-031-62656-2_27

Nouali, O., Moussaoui, S., & Derder, A. (2015). A BLE-based data collection system for IoT. In *Proceedings of the 2015 First International Conference on New Technologies of Information and Communication (NTIC)* (pp. 1–5). Mila, Algeria. https://doi.org/DOI: 10.1109/NTIC.2015.7368783

Nowacki, L. (2024). The old king is dead, long live the algorithmic king – the decline of Facebook and the rise of TikTok – comparative study of algorithmic design of social media platforms. *International Social Science Journal, 74*(254), 1325–1338. Advance online publication. DOI: 10.1111/issj.12513

Nurcahyo, A. C. (2024). *Digital Literacy for Christian Students in Strengthening Faith within the Digital Realm.* AMERTA MEDIA.

Nwaimo, N. C. S., Adegbola, N. E., & Adegbola, N. M. D.Chioma Susan NwaimoAyodeji Enoch AdegbolaMayokun Daniel Adegbola. (2024). Data-driven strategies for enhancing user engagement in digital platforms. *International Journal of Management & Entrepreneurship Research, 6*(6), 1854–1868. DOI: 10.51594/ijmer.v6i6.1170

Odejide, N. O. A., & Edunjobi, N. T. E. (2024). AI IN PROJECT MANAGEMENT: EXPLORING THEORETICAL MODELS FOR DECISION-MAKING AND RISK MANAGEMENT. *Engineering Science & Technology Journal, 5*(3), 1072–1085. DOI: 10.51594/estj.v5i3.959

Ofulue, J., & Benyoucef, M. (2022). *Data monetization: insights from a technology-enabled literature review and research agenda.* Management Review Quarterly., DOI: 10.1007/s11301-022-00309-1

Ojeniyi, J. A., Edward, E. O., & Abdulhamid, S. M. (2019). Security risk analysis in online banking transactions: Using diamond bank as a case study. *International Journal of Education and Management Engineering, 9*(2), 1–14. DOI: 10.5815/ijeme.2019.02.01

Oke, A. E., Aliu, J., Mwanaumo, E. M. M., Abayomi, T., & Kahanji, C. (2023). Leveraging gamification to enhance productivity and employee engagement in the Nigerian construction industry. *Built Environment Project and Asset Management, 13*(6), 813–829. DOI: 10.1108/BEPAM-04-2023-0064

Okeyo, M. (2024). *Customer-centric data strategies to drive revenue growth.* eCommons@AKU. https://ecommons.aku.edu/theses_dissertations/2310/

Okumuş, M. S. (2022). *THE AUDIENCE STRUGGLE BETWEEN TELEVISION, CINEMA, AND OTT PLATFORMS: THE CASE OF NETFLIX AND DISNEY+ IN TURKEY. DergiPark.* Istanbul University., DOI: 10.7456/11301100/008

Olabanji, S. O., Olaniyi, O. O., Adigwe, C. S., Okunleye, O. J., & Oladoyinbo, T. O. (2024). AI for Identity and Access Management (IAM) in the Cloud: Exploring the Potential of Artificial Intelligence to Improve User Authentication, Authorization, and Access Control within Cloud-Based Systems. SSRN *Electronic Journal.* DOI: 10.2139/ssrn.4706726

Olaoye, F. "Quantum Computing for Cryptographic Security." [Online]. Available: https://www.researchgate.net/publication/384678504

Oliveira, J., Santin, A., Viegas, E., & Horchulhack, P. (2024). A non-interactive One-Time Password-Based method to enhance the vault security. In *Lecture notes on data engineering and communications technologies* (pp. 201–213). DOI: 10.1007/978-3-031-57916-5_18

Ong, E. B. H., & Samson, B. P. V. (2024). Exploring design opportunities for improved self-motivation in self-tracking and health goal achievement. *Proceedings of the ACM on Human-Computer Interaction, 8*(MHCI), 1–17. DOI: 10.1145/3676501

Onyebuchi, A., Matthew, U. O., Kazaure, J. S., Okafor, N. U., Okey, O. D., Okochi, P. I., Taiwo, J. F., & Matthew, A. O. (2022). Business demand for a cloud enterprise data warehouse in electronic healthcare computing: Issues and developments in e-healthcare cloud computing. [IJCAC]. *International Journal of Cloud Applications and Computing*, *12*(1), 1–22. DOI: 10.4018/IJCAC.297098

Oravec, J. A. (2015). Gamification and multigamification in the workplace: Expanding the ludic dimensions of work and challenging the work/play dichotomy. *Cyberpsychology (Brno)*, *9*(3). Advance online publication. DOI: 10.5817/CP2015-3-6

Orji, R., Alslaity, A., & Chan, G. (2023). Towards understanding the mechanism through which reward and punishment motivate or demotivate behaviours. *Behaviour & Information Technology*, *43*(6), 1042–1066. DOI: 10.1080/0144929X.2023.2196582

Overbeck, R., & Sendrier, N. (n.d.). Code-based cryptography. In *Post-Quantum Cryptography* (pp. 95–145). Springer Berlin Heidelberg., DOI: 10.1007/978-3-540-88702-7_4

Pandey, M., Litoriya, R., & Pandey, P. (2020). Applicability of machine learning methods on mobile app effort estimation: Validation and performance evaluation. *International Journal of Software Engineering and Knowledge Engineering*, *30*(01), 23–41. DOI: 10.1142/S0218194020500023

Paschmann, J. W., Bruno, H. A., Van Heerde, H. J., Völckner, F., & Klein, K. (2024). EXPRESS: Driving Mobile App User Engagement through Gamification. *JMR, Journal of Marketing Research.* Advance online publication. DOI: 10.1177/00222437241275927

Patil, D., Rane, N. L., & Rane, J. (2024). The future of customer loyalty: How ChatGPT and generative artificial intelligence are transforming customer engagement, personalization, and satisfaction. In *Science DIrect.* DOI: 10.70593/978-81-981367-8-7_2

Paudel, R., & Al-Ameen, M. N. (2024). Priming through Persuasion: Towards Secure Password Behavior. *Proceedings of the ACM on Human-Computer Interaction, 8*(CSCW1), 1–27. DOI: 10.1145/3637387

Payares, E., & Martinez-Santos, J. C. (2021). Quantum machine learning for intrusion detection of distributed denial of service attacks: A comparative overview. In *Proceedings of SPIE Quantum Computing, Communication, and Simulation* (p. 47). DOI: 10.1117/12.2593297

Pekpak, N., & Altunalan, T. (2024). Relationship Between Screen Time with Posture and Motor Coordination in Childhood: A Descriptive Study. *Turkiye Klinikleri Journal of Pediatrics, 33*(2), 41–47. DOI: 10.5336/pediatr.2024-101303

Petrou, S., Okonji, P., Olufemi, O., Kpokiri, E., Wood, C., Luedtke, S., Birjovanu, G., Aldosery, A., Ogunsola, F., & Kostkova, P. (2024). Gamified Antimicrobial Decision Support App (GADSA)-Is gamification more effective user engagement than group chat? *2024 IEEE 12th International Conference on Serious Games and Applications for Health (SeGAH)*, 1–7. DOI: 10.1109/SeGAH61285.2024.10639568

Petsios, T., Zhao, J., Keromytis, A. D., & Jana, S. "SlowFuzz," in *Proceedings of the 2017 ACM SIGSAC Conference on Computer and Communications Security*, New York, NY, USA: ACM, Oct. 2017, pp. 2155–2168. DOI: 10.1145/3133956.3134073

Pham, H., Nhuan, N. D. O., & Huynh-Tuong, N. (2023). A Fine-grained Access Control Model with Enhanced Flexibility and On-chain Policy Execution for IoT Systems. *International Journal of Advanced Computer Science and Applications, 14*(6). Advance online publication. DOI: 10.14569/IJACSA.2023.0140609

Pienta, D., Thatcher, J. B., Wright, R. T., & Roth, P. L. (2024). An empirical investigation of the unintended consequences of vulnerability assessments leading to betrayal. *Journal of the Association for Information Systems, 25*(4), 1079–1116. DOI: 10.17705/1jais.00875

Pizzo, A. D., Fornaciari, C. J., Coyle, P., Varzgani, N., & Pierce, M. (2024). Opening the metaverse: The next frontier in sport and leisure. *Leisure Sciences*, •••, 1–19. DOI: 10.1080/01490400.2024.2407023

Pombo, N., & Santos, H. (2023). Lessons Learned from the Development of a Computerised Badge-based Reward Tool for Student Engagement in Learning Activities. *2023 IEEE World Engineering Education Conference (EDUNINE)*. DOI: 10.1109/EDUNINE57531.2023.10102849

Prasad, A. (2024). A Comparative Study of Passwordless Authentication. *TechRxiv*. DOI: 10.36227/techrxiv.171560547.71979752/v1

Pratama, S. F. (2024). Temporal Patterns in user Conversions: Investigating the impact of ad scheduling in digital marketing. *Journal of Digital Market and Digital Currency.*, *1*(2), 165–182. DOI: 10.47738/jdmdc.v1i2.10

Predescu, A., & Mocanu, M. (2024). Gamification in Real-World applications: interactive maps and augmented reality. In *IntechOpen eBooks*. DOI: 10.5772/intechopen.1004870

Prinakaa, S., Bavanika, V., Sanjana, S., Srinivasan, S., & Sarasvathi, V. (2024, April). A Real-Time Approach to Detecting API Abuses Based on Behavioral Patterns. In 2024 8th International Conference on Cryptography, Security and Privacy (CSP) (pp. 24-28). IEEE.

Prümmer, J. (2024). The Role of Cognition in Developing Successful Cybersecurity Training Programs – Passive vs. Active Engagement. In *Lecture notes in computer science* (pp. 185–199). DOI: 10.1007/978-3-031-61572-6_13

Punde, P. A., Jadhav, M. E., & Manza, R. R. (2017, October). A study of eye tracking technology and its applications. In *2017 1st International Conference on Intelligent Systems and Information Management (ICISIM)* (pp. 86-90). IEEE. DOI: 10.1109/ICISIM.2017.8122153

Putra, E. A. (2024, November 23). *The concept of Gharar in Purchasing Mobile Legends Virtual Items: A study of the clarity of objects and transactions*. Putra | Saqifah: Jurnal Hukum Ekonomi Syariah. https://journals.fasya.uinib.org/index.php/saqifah/article/view/638

Qamar, A., Karim, A., & Chang, V. (2019). Mobile malware attacks: Review, taxonomy & future directions. *Future Generation Computer Systems*, *97*, 887–909. DOI: 10.1016/j.future.2019.03.007

Rabbi, M., Pfammatter, A., Zhang, M., Spring, B., & Choudhury, T. (2015). Automated Personalized feedback for physical activity and dietary behavior change with mobile phones: A randomized controlled trial on adults. *JMIR mHealth and uHealth*, *3*(2), e42. DOI: 10.2196/mhealth.4160 PMID: 25977197

Rabhan, B. (2013). Convert Every Click: Make More Money Online with Holistic Conversion Rate Optimization. In *Wiley Publishing eBooks*. https://dl.acm.org/citation.cfm?id=2566853

Radivojevic, K., Clark, N., Klempay, A., & Brenner, P. (2024). Defending novice user privacy: An evaluation of default web browser configurations. *Computers & Security*, *140*, 103784. DOI: 10.1016/j.cose.2024.103784

Ramadani, R. (2024). The impact of technology use on Young people: A case study of social media and internet usage. *Asian Journal of Research in Computer Science*, *17*(8), 13–23. DOI: 10.9734/ajrcos/2024/v17i7486

Rana, P., Batra, I., Malik, A., Ra, I.-H., Lee, O.-S., & Hosen, A. S. M. S. (2024). Efficacious novel intrusion detection system for cloud computing environment. *IEEE Access : Practical Innovations, Open Solutions*, *12*, 99223–99239. DOI: 10.1109/ACCESS.2024.3424528

Rani, P., Verma, S., Yadav, S. P., Rai, B. K., Naruka, M. S., & Kumar, D. (2022). Simulation of the lightweight blockchain technique based on privacy and security for healthcare data for the cloud system. [IJEHMC]. *International Journal of E-Health and Medical Communications*, *13*(4), 1–15. DOI: 10.4018/IJEHMC.309436

Raphael, M. M., Kanta, A., Seebonn, R., Dürmuth, M., & Cobb, C. (2024). *Batman Hacked my Password: A {Subtitle-Based} analysis of password depiction in movies*. https://www.usenix.org/conference/soups2024/presentation/raphael

Ravichandran, K., Virgin, B. A., Patil, S., Fatma, G., Rengarajan, M., & Bala, B. K. (2024). Gamifying Language Learning: Applying Augmented Reality and Gamification Strategies for Enhanced English Language Acquisition. *2024 Third International Conference on Smart Technologies and Systems for Next Generation Computing (ICSTSN)*, 1–6. DOI: 10.1109/ICSTSN61422.2024.10670787

Ravichandran, N., Tewaraja, T., Rajasegaran, V., Kumar, S. S., Gunasekar, S. K. L., & Sindiramutty, S. R. (2024). Comprehensive Review Analysis and Countermeasures for Cybersecurity Threats: DDoS, Ransomware, and Trojan Horse Attacks. *Preprints.org*. DOI: 10.20944/preprints202409.1369.v1

Ravikumar, C., Batra, I., & Malik, A. (2022). A comparative analysis on blockchain technology considering security breaches. In *Lecture Notes in Networks and Systems* (Vol. 376, pp. 555–565).

Ravikumar, C., Kumar, K. H., Sathish, N., Suhasini, S., & Nimmala, S. (2024). A comprehensive analysis of machine learning methods for bug prediction in software development. *Lecture Notes in Electrical Engineering, 1273*, 929–935. DOI: 10.1007/978-981-97-8031-0_99

Ravikumar, C., Naresh, B., Prasanna, P. L., Goud, E. A., & Prasad, P. R. (2024). Exploring machine learning algorithms for robust cyber threat detection and classification: A comprehensive evaluation. In *2024 International Conference on Intelligent Systems for Cybersecurity (ISCS).*

Rege, I., & Dimitrova, N. (2024). *How does Ad Avoidance on phones affect Consumer Decision-Making?: An exploratory study on the effect thatad avoidance has on customer decision-making*. DIVA. https://www.diva-portal.org/smash/record.jsf?pid=diva2%3A1864770&dswid=-8184

Rendell, R., Pinheiro, M., Wang, B., McKay, F., Ewen, A., Carnegie, C., Tikomaidelana, E., Fattah, Z., & Hassett, L. (2024). Digital Apps to Improve Mobility in Adults with Neurological Conditions: A Health App-Focused Systematic Review. *Health Care, 12*(9), 929. DOI: 10.3390/healthcare12090929 PMID: 38727486

Rezaee, K., Rezakhani, S. M., Khosravi, M. R., & Moghimi, M. K. (2021). A survey on deep learning-based real-time crowd anomaly detection for secure distributed video surveillance. *Personal and Ubiquitous Computing*. Advance online publication. DOI: 10.1007/s00779-021-01586-5

Ringborg, L., & Ohlsson, I. (2024). *The influence of TikTok : Understanding the Influence of TikTok UGC on Gen Z's Impulsive Buying Behavior regarding Beauty Products.* DIVA. https://www.diva-portal.org/smash/record.jsf?pid=diva2%3A1862715&dswid=72

Rinpoche, D. K. (2024). *The free mind: Finding Clarity in a Digitally Distracted World.* Shambhala Publications.

Rodrik, D., & Sandhu, R. (2024). *Servicing Development: Productive upgrading of Labor-Absorbing services in developing economies.* DOI: 10.3386/w32738

Rooney, M. J., Levy, Y., Li, W., & Kumar, A. (2024). Comparing experts' and users' perspectives on the use of password workarounds and the risk of data breaches. *Information and Computer Security*. Advance online publication. DOI: 10.1108/ICS-05-2024-0116

Ros, R., Bjarnason, E., & Runeson, P. (2023). A theory of factors affecting continuous experimentation (FACE). *Empirical Software Engineering*, *29*(1), 21. Advance online publication. DOI: 10.1007/s10664-023-10358-z

Rupesh, V. (2023, September). Quantum Secure Communication: A Comprehensive Literary Analysis with In-Depth Insights. *Bioscience Biotechnology Research Communications*, *16*(3), 171–177. DOI: 10.21786/bbrc/16.3.7

Rusin, N., & Szandała, T. (2024). The power of peer recognition points: Does it really boost employee engagement? *Strategic HR Review*. Advance online publication. DOI: 10.1108/SHR-06-2024-0040

Rutledge, P., Garner, M. S., & Collins, H. N. (2024). Exercising empowerment: Mediated fitness technology fostered social connections and personal growth during social isolation. *Psychology of Popular Media*. DOI: 10.1037/ppm0000517

Ryu, S. (2024). *Media and entertainment industry management: How to Integrate Business and Management with Creativity and Imagination*. Taylor & Francis. DOI: 10.4324/9781003271222

Saadi, Z. M., Sadiq, A. T., Akif, O. Z., & Farhan, A. K. (2024a). A Survey: Security Vulnerabilities and Protective Strategies for graphical Passwords. *Electronics (Basel)*, *13*(15), 3042. DOI: 10.3390/electronics13153042

Salem, A., Sharieh, A., & Jabri, R. (2022). Online user authentication System using keystroke dynamics. *Journal of Computer Security*, *31*(3), 185–215. DOI: 10.3233/JCS-210081

Sama, N. U., Zen, K., Humayun, M., Jhanjhi, N. Z., & Rahman, A. U. (2022). Security in wireless body sensor network: A multivocal literature study. *Applied System Innovation*, *5*(4), 79. DOI: 10.3390/asi5040079

Sanchez, E., Van Oostendorp, H., Fijnheer, J. D., & Lavoué, E. (2020). Gamification. In *Springer eBooks* (pp. 816–827). DOI: 10.1007/978-3-030-10576-1_38

Saqib, M., & Moon, A. H. (2022). A systematic security assessment and review of internet of things in the context of authentication. *Computers & Security*, *125*, 103053. DOI: 10.1016/j.cose.2022.103053

Sarangdhar, V., Awate, S., & Mudambi, R. (2024). Business model innovations in high-velocity environments. *Journal of Business Research*, *183*, 114864. DOI: 10.1016/j.jbusres.2024.114864

Sarfaraz, A., Chakrabortty, R. K., & Essam, D. L. (2023). AccessChain: An access control framework to protect data access in blockchain enabled supply chain. *Future Generation Computer Systems, 148*, 380–394. DOI: 10.1016/j.future.2023.06.009

Saxena, U. R., & Alam, T. (2023). Role-based access using partial homomorphic encryption for securing cloud data. *International Journal of System Assurance Engineering and Management, 14*(3), 950–966. DOI: 10.1007/s13198-023-01896-2

Scanlon, D., Coulter, M., Baker, K., & Tannehill, D. (2024). The enactment of the Socially-Just Teaching Personal and Social Responsibility (SJ-TPSR) approach in physical education teacher education: Teacher educators' and pre-service teachers' perspectives. *Physical Education and Sport Pedagogy*, •••, 1–17. DOI: 10.1080/17408989.2024.2319071

Scherr, S. A., Ter-Tovmasyan, M., Neugebauer, F., Hupp, S., & Elberzhager, F. (2022). The way it made me feel-Creating and evaluating an in-app feedback tool for mobile apps. *J. Ubiquitous Syst. Pervasive Networks, 17*(1), 27–34.

Schluchter, T., Ahmed, N., & Huang, E. (2024). "It Actually Affected My Relationship": A Qualitative Analysis of Affordances and Attitudes Towards Mental Health Content on TikTok. *ACM, 5*, 491–507. DOI: 10.1145/3670653.3677501

Sellgren, J. (2024). *Beyond the purchase : Exploring the dynamics of microtransactions among Swedish gamers*. DIVA. https://www.diva-portal.org/smash/record.jsf?pid=diva2%3A1867555&dswid=-7330

Seng, Y. J., Cen, T. Y., Raslan, M. H. B. M., Subramaniam, M. R., Xin, L. Y., Kin, S. J., Long, M. S., & Sindiramutty, S. R. (2024). In-Depth Analysis and Countermeasures for Ransomware Attacks: Case Studies and Recommendations. *Preprints.org*. DOI: 10.20944/preprints202408.2261.v1

Seraj, S., Sadiq, A. S., Kaiwartya, O., Aljaidi, M., Konios, A., Ali, M., & Abazeed, M. (2024). MoBShield: A Novel XML Approach for SecuringMobile Banking. *Computers, Materials & Continua, 79*(2), 2123–2149. DOI: 10.32604/cmc.2024.048914

Setiyawati, N., & Bangkalang, D. H. (2022, September). The comparison of evaluation on user experience and usability of mobile banking applications using user experience questionnaire and system usability scale. In *Proceedings* (Vol. 82, No. 1, p. 87). MDPI. DOI: 10.3390/proceedings2022082087

Shah, I. A., Jhanjhi, N. Z., & Laraib, A. (2022). Cybersecurity and blockchain usage in contemporary business. In *Advances in information security, privacy, and ethics book series* (pp. 49–64). DOI: 10.4018/978-1-6684-5284-4.ch003

Shah, I. A., Jhanjhi, N., & Brohi, S. N. (2024). Proposing model for classification of malicious SQLi code using machine learning approach. In *2024 1st International Conference on Innovative Engineering Sciences and Technological Research (ICI-ESTR)* (pp. 1-6). DOI: 10.1109/ICIESTR60916.2024.10798230

Shah, I. A., Jhanjhi, N. Z., & Brohi, S. N. (2025). Machine learning models for detecting software vulnerabilities. In *Generative AI for Web Engineering Models* (pp. 1–40). IGI Global.

Shah, S. A. A., Ullah, A., Subhan, F., Jhanjhi, N. Z., Masud, M., & Alqhatani, A. (2024). Truth discovery for mobile workers in edge-assisted mobile crowdsensing. *ICT Express*, *10*(5), 1087–1093. DOI: 10.1016/j.icte.2024.06.007

Sharma, R., Singh, A., Kavita, N., Jhanjhi, N. Z., Masud, M., Jaha, E. S., & Verma, S. (2021). Plant disease diagnosis and image classification using deep learning. *Computers, Materials & Continua/Computers. Materials & Continua (Print)*, *71*(2), 2125–2140. DOI: 10.32604/cmc.2022.020017

Sharma, W., Lim, W. M., Kumar, S., Verma, A., & Kumra, R. (2023). Game on! A state-of-the-art overview of doing business with gamification. *Technological Forecasting and Social Change*, *198*, 122988. DOI: 10.1016/j.techfore.2023.122988

Sheil, A., Acar, G., Schraffenberger, H., Gellert, R., & Malone, D. (2024). *Staying at the Roach Motel: Cross-Country Analysis of Manipulative Subscription and Cancellation Flows*. ACM., DOI: 10.1145/3613904.3642881

Sheng, L. T., Azman, A. B., Khan, N. A., Sindiramutty, S. R., Singh, S., & Verma, S. (2024). Reinforcement learning-based traffic signal control. *2024 International Conference on Emerging Trends in Networks and Computer Communications (ET-NCC)*, 792–798. DOI: 10.1109/ETNCC63262.2024.10767501

Shijo, P. V., & Salim, A. (2015). Integrated Static and Dynamic Analysis for Malware Detection. *Procedia Computer Science*, *46*, 804–811. DOI: 10.1016/j.procs.2015.02.149

Shor, P. W., & Preskill, J. (2000, July). Simple proof of security of the BB84 quantum key distribution protocol. *Physical Review Letters*, *85*(2), 441–444. DOI: 10.1103/PhysRevLett.85.441 PMID: 10991303

Shortt, M., Tilak, S., Kuznetcova, I., Martens, B., & Akinkuolie, B. (2021). Gamification in mobile-assisted language learning: A systematic review of Duolingo literature from public release of 2012 to early 2020. *Computer Assisted Language Learning*, *36*(3), 517–554. DOI: 10.1080/09588221.2021.1933540

Shu-Chen, C. (2023). From freemium to premium app services: The Expectation confirmation model and Two-Factor Theory. *International Journal of Marketing Communication and New Media, 11*(21). Advance online publication. DOI: 10.54663/2182-9306.2023.v11.n21.112-131

Shukla, R. S., Ghuse, E. A., Diwan, T., Tembhurne, J. V., & Sahare, P. (2024). GA-MPG: Efficient genetic algorithm for improvised mobile plan generation. *Journal of Ambient Intelligence and Humanized Computing, 15*(10), 3675–3691. DOI: 10.1007/s12652-024-04846-3

Sindiramutty, S. R., Jhanjhi, N. Z., Tan, C. E., Khan, N. A., Shah, B., & Manchuri, A. R. (2024). Cybersecurity measures for logistics industry. In *Advances in information security, privacy, and ethics book series* (pp. 1–58). DOI: 10.4018/979-8-3693-3816-2.ch001

Sindiramutty, S. R., Jhanjhi, N. Z., Tan, C. E., Khan, N. A., Shah, B., Yun, K. J., Ray, S. K., Jazri, H., & Hussain, M. (2024). Future trends and emerging threats in drone cybersecurity. In *Advances in information security, privacy, and ethics book series* (pp. 148–195). DOI: 10.4018/979-8-3693-0774-8.ch007

Sindiramutty, S. R., Jhanjhi, N. Z., Tan, C. E., Yun, K. J., Manchuri, A. R., Ashraf, H., Murugesan, R. K., Tee, W. J., & Hussain, M. (2024). Data security and privacy concerns in drone operations. In *Advances in information security, privacy, and ethics book series* (pp. 236–290). DOI: 10.4018/979-8-3693-0774-8.ch010

Sindiramutty, S. R., Jhanjhi, N., Tan, C. E., Lau, S. P., Muniandy, L., Gharib, A. H., Ashraf, H., & Murugesan, R. K. (2024). Industry 4.0. In *Advances in logistics, operations, and management science book series* (pp. 342–405). DOI: 10.4018/979-8-3693-1363-3.ch013

Sindiramutty, S. R., Prabagaran, K. R. V., Akbar, R., Hussain, M., & Malik, N. A. (2024). Overview of Generative AI techniques for cybersecurity. In *Advances in information security, privacy, and ethics book series* (pp. 1–52). DOI: 10.4018/979-8-3693-5415-5.ch001

Sindiramutty, S. R., Prabagaran, K. R. V., Jhanjhi, N. Z., Ghazanfar, M. A., Malik, N. A., & Soomro, T. R. (2024). Security Considerations in Generative AI for web Applications. In *Advances in information security, privacy, and ethics book series* (pp. 281–332). DOI: 10.4018/979-8-3693-5415-5.ch009

Sindiramutty, S. R., Prabagaran, K. R. V., Jhanjhi, N. Z., Murugesan, R. K., Brohi, S. N., & Masud, M. (2024). Generative AI in network security and intrusion detection. In *Advances in information security, privacy, and ethics book series* (pp. 77–124). DOI: 10.4018/979-8-3693-5415-5.ch003

Sindiramutty, S. R., Prabagaran, K. R. V., Jhanjhi, N. Z., Murugesan, R. K., Brohi, S. N., & Wei, G. W. (2024). Generative AI for threat intelligence and information sharing. In *Advances in digital crime, forensics, and cyber terrorism book series* (pp. 191–234). DOI: 10.4018/979-8-3693-8944-7.ch006

Sindiramutty, S. R., Prabagaran, K. R. V., Jhanjhi, N. Z., Murugesan, R. K., Malik, N. A., & Hussain, M. (2024). Ethics and transparency in secure web model generation. In *Advances in information security, privacy, and ethics book series* (pp. 411–464). DOI: 10.4018/979-8-3693-5415-5.ch012

Sindiramutty, S. R., Tan, C. E., & Wei, G. W. (2024). Eyes in the sky. In *Advances in information security, privacy, and ethics book series* (pp. 405–451). DOI: 10.4018/979-8-3693-0774-8.ch017

Sindiramutty, S. R., Tan, C. E., Shah, B., Khan, N. A., Gharib, A. H., Manchuri, A. R., Muniandy, L., Ray, S. K., & Jazri, H. (2024). Ethical considerations in drone cybersecurity. In *Advances in information security, privacy, and ethics book series* (pp. 42–87). DOI: 10.4018/979-8-3693-0774-8.ch003

Singhal, V., Jain, S. S., Anand, D., Singh, A., Verma, S., Kavita, N., Rodrigues, J. J. P. C., Jhanjhi, N. Z., Ghosh, U., Jo, O., & Iwendi, C. (2020). Artificial Intelligence Enabled Road Vehicle-Train Collision Risk Assessment Framework for Unmanned railway level crossings. *IEEE Access : Practical Innovations, Open Solutions, 8*, 113790–113806. DOI: 10.1109/ACCESS.2020.3002416

Singh, Y., & Milan, R. (2024). Utilitarian and hedonic values of gamification and their influence on brand engagement, loyalty, trust and WoM. *Entertainment Computing, 100868*. Advance online publication. DOI: 10.1016/j.entcom.2024.100868

Siricharoen, W. V. (2023). Improving user experience (UX) by applying (Interactive) infographic in the human computer interaction context. *Mobile Networks and Applications*. Advance online publication. DOI: 10.1007/s11036-023-02179-7

Slamet, T., Brush, T., & Kwon, K. (2024). *The effects of competition in gamified online discussions on learners' behavioral and cognitive engagement*. Technology Knowledge and Learning., DOI: 10.1007/s10758-024-09740-7

Smmarwar, S. K., Gupta, G. P., & Kumar, S. (2024). Android malware detection and identification frameworks by leveraging the machine and deep learning techniques: A comprehensive review. *Telematics and Informatics Reports, 14*, 100130. DOI: 10.1016/j.teler.2024.100130

Somani, V., & Trivedi, M. C. (2019). Utilizing cloud computing for stronger healthcare data security. *Journal of Advanced Research in Dynamical and Control Systems*, *11*(11), 60–69. DOI: 10.5373/JARDCS/V11I11/20193169

Sood, N. “Cryptography in Post Quantum Computing Era”, doi: DOI: 10.13140/RG.2.2.19691.92964

Sotos-Martínez, V. J., Ferriz-Valero, A., García-Martínez, S., & Tortosa-Martínez, J. (2022). The effects of gamification on the motivation and basic psychological needs of secondary school physical education students. *Physical Education and Sport Pedagogy*, •••, 1–17. DOI: 10.1080/17408989.2022.2039611

Sprengers, M., & van Haaster, J. (2016). Organization of #operations. In *Cyber Guerilla* (pp. 41–109). Elsevier. DOI: 10.1016/B978-0-12-805197-9.00003-6

Suhod, M. S. M., Ngah, A. H., & Rahi, S. (2024). Unravelling the continue of subscribing Spotify Premium among university students: The extended UTAUT 2 model. *International Journal of Business Information Systems*, *46*(2), 217–237. DOI: 10.1504/IJBIS.2024.139908

Sukmana, Y., & Rosmansyah, Y. (2021). The use of Cloud FireStore for handling real-time data updates: An Empirical study of gamified Online quiz. *2021 2nd International Conference on Electronics, Communications and Information Technology (CECIT)*. DOI: 10.1109/CECIT53797.2021.00220

Temel, E. (2024, September 27). *Qualitative Analysis of Consumer’s Responses to Mobile Push Notifications and Its Relationship with Impulsive Purchasing*. https://dergipark.org.tr/en/pub/jcsci/issue/87400/1557272

Thakur, N., & Han, C. Y. (2021). Multimodal approaches for indoor localization for ambient assisted living in smart homes. *Information (Basel)*, *12*(3), 114. DOI: 10.3390/info12030114

Tirole, J. (2024). Competition and industrial policy in the 21st century. *Oxford Open Economics*, *3*(Supplement_1), i983–i1001. DOI: 10.1093/ooec/odad080

Tiwary, P., Pandey, A., Kumar, S., & Youssef, M. (2021). Novel differential r-vectors for localization in IoT networks. *IEEE Sensors Letters*, *5*(6), 1–4. DOI: 10.1109/LSENS.2021.3083500

Tomic, S., Beko, M., & Dinis, R. (2022). *RSS-AoA-based target localization and tracking in wireless sensor networks*. CRC Press. DOI: 10.1201/9781003339335

Tong, F., & Yan, Z. (2017). A hybrid approach of mobile malware detection in Android. *Journal of Parallel and Distributed Computing, 103*, 22–31. DOI: 10.1016/j.jpdc.2016.10.012

Tongpaeng, Y., Nobnop, R., Wongwan, N., Homla, P., Intawong, K., & Puritat, K. (2024). Comparison of gamified and non-gamified mixed reality in enhancing museum visitor engagement, motivation, and learning outcome. *Journal of Heritage Tourism, 19*(6), 1–30. DOI: 10.1080/1743873X.2024.2351852

Trampert, L., Stock, B., & Roth, S. (2023). *Honey, I Cached our Security Tokens Re-usage of Security Tokens in the Wild.* ACM., DOI: 10.1145/3607199.3607223

Trinh, M. P., Chico, R. J., & Reed, R. M. (2023). How fun Overcame Fear: The gamification of a Graduate-Level Statistics course. *Journal of Management Education, 48*(4), 735–776. DOI: 10.1177/10525629231181120

Troussas, C., Krouska, A., & Virvou, M. (2017). Reinforcement theory combined with a badge system to foster student's performance in e-learning environments. *2017 8th International Conference on Information* [IISA]. *Intelligent Systems with Applications, 38*, 1–6. DOI: 10.1109/IISA.2017.8316421

Tsai, C. H., Liou, D. K., & Lee, H. L. (2024). Blockchain-supported online banking scheme. *Egyptian Informatics Journal, 27*, 100516. DOI: 10.1016/j.eij.2024.100516

Tsikerdekis, M., Waldron, S., & Emanuelson, A. (2021). Network Anomaly Detection Using Exponential Random Graph Models and Autoregressive Moving Average. *IEEE Access : Practical Innovations, Open Solutions, 9*, 134530–134542. DOI: 10.1109/ACCESS.2021.3116575

Upadrista, V., Nazir, S., & Tianfield, H. (2023). Secure data sharing with blockchain for remote health monitoring applications: A review. *Journal of Reliable Intelligent Environments, 9*(3), 349–368. DOI: 10.1007/s40860-023-00204-w PMID: 37359293

Ur, B., Bees, J., Segreti, S. M., Bauer, L., Christin, N., & Cranor, L. F. (2016). *Do Users' Perceptions of Password Security Match Reality?* ACM., DOI: 10.1145/2858036.2858546

Vaitone, N. V. (2025, May 27). -., Kirse, S., Sakalauske, K. a.-., Dikcius, V., & Zimaitis, I. (2024). The usefulness of gamification for enhancing customer loyalty to small e-tailers. *EuroMed Journal of Business, 20*(2), 435–452. Advance online publication. DOI: 10.1108/EMJB-09-2023-0240

Varadarajan, M. N., & Seeni, S. K. (2024). INNOVATIVE DIGITAL OWNERSHIP AND COLLECTIBLES VIA PROOF OF STAKE (POS) AND NON-FUNGIBLE TOKENS (NFTS). *International Journal of Advances in Signal and Image Sciences*, *10*(1), 22–34. DOI: 10.29284/IJASIS.10.1.2024.22-34

Varma, P. S., & Anand, V. (2021). Random forest learning based indoor localization as an IoT service for smart buildings. *Wireless Personal Communications*, *117*(4), 3209–3227. DOI: 10.1007/s11277-020-07977-w

Vashisht, D. (2023). Engaging and Entertaining customers: Gamification in interactive marketing. In *Springer eBooks* (pp. 807–835). DOI: 10.1007/978-3-031-14961-0_35

Verma, V., Malik, A., & Batra, I. (2023). Assessing the Impact and Performance of Malware Detection Systems. *2023 10th IEEE Uttar Pradesh Section International Conference on Electrical, Electronics and Computer Engineering (UPCON)*, 106–109. DOI: 10.1109/UPCON59197.2023.10434443

Vinay, M. D., Kumar, M. H., Hemanth, B., & Tomar, D. S. (2024). The Demonstration of Android Vulnerabilities. *2024 IEEE International Students' Conference on Electrical, Electronics and Computer Science (SCEECS)*, *11*, 1–10. DOI: 10.1109/SCEECS61402.2024.10482193

Waheed, A., Seegolam, B., Jowaheer, M. F., Sze, C. L. X., Hua, E. T. F., & Sindiramutty, S. R. (2024). Zero-Day Exploits in Cybersecurity: Case Studies and Countermeasure. *preprints.org*. DOI: 10.20944/preprints202407.2338.v1

Wang, L., Liu, X., Lei, T., Song, W., Guo, S., & Ren, P. (2024). Security research for Android remote assistance apps. In *Lecture notes in computer science* (pp. 256–276). DOI: 10.1007/978-981-97-5101-3_14

Wang, S., Asif, M., Shahzad, M. F., & Ashfaq, M. (2024). Data privacy and cybersecurity challenges in the digital transformation of the banking sector. *Computers & Security*, *147*, 104051. DOI: 10.1016/j.cose.2024.104051

Wang, S., Chen, Z., Yan, Q., Yang, B., Peng, L., & Jia, Z. (2019). A mobile malware detection method using behavior features in network traffic. *Journal of Network and Computer Applications*, *133*, 15–25. DOI: 10.1016/j.jnca.2018.12.014

Wang, W., & Sari, M. K. (2023). Examining the effect of the Task-Technology fit of game mechanisms on learning outcomes in online gamification platforms. *Journal of Educational Computing Research*, *61*(8), 32–59. DOI: 10.1177/07356331231187285

Wan, L., Zhang, M., Sun, L., & Wang, X. (2021). Machine learning empowered IoT for intelligent vehicle location in smart cities. *ACM Transactions on Internet Technology*, *21*(3), 1–25. DOI: 10.1145/3448612

Wear, S. (2018). *Burp Suite Cookbook: Practical recipes to help you master web penetration testing with Burp Suite*. Packt Publishing Ltd.

Weiqi, X., Hooi, S. T. C., Sindiramutty, S. R. L., Asirvatham, D. L., Kumar, D., & Verma, S. (2024). Surface Anomaly Detection Using Machine Learning Technique. *2024 International Conference on Emerging Trends in Networks and Computer Communications (ETNCC)*, 1–7. DOI: 10.1109/ETNCC63262.2024.10767562

Wen, B. O. T., Syahriza, N., Xian, N. C. W., Wei, N. G., Shen, T. Z., Hin, Y. Z., Sindiramutty, S. R., & Nicole, T. Y. F. (2023). Detecting cyber threats with a Graph-Based NIDPS. In *Advances in logistics, operations, and management science book series* (pp. 36–74). DOI: 10.4018/978-1-6684-7625-3.ch002

What is monetization? | Adjust. (2024). https://www.adjust.com/glossary/app-monetization/

Wodo, W., Stygar, D., & Blaskiewicz, P. (2021, July). Security Issues of Electronic and Mobile Banking. In *SECRYPT* (pp. 631-638).

Xi, N., & Hamari, J. (2019). Does gamification satisfy needs? A study on the relationship between gamification features and intrinsic need satisfaction. *International Journal of Information Management*, *46*, 210–221. DOI: 10.1016/j.ijinfomgt.2018.12.002

Xu, W., Jha, S., & Hu, W. (2019, August). LoRa-Key: Secure key generation system for LoRa-based network. *IEEE Internet of Things Journal*, *6*(4), 6404–6416. DOI: 10.1109/JIOT.2018.2888553

Yan, C., Yang, Z., & Dai, X. (2023). Subscriptions versus one-off purchases: The impact of fee-charging models on consumer privacy concerns. *International Journal of Bank Marketing*, *41*(4), 835–859. DOI: 10.1108/IJBM-06-2022-0271

Yang, Y., Wang, J., Liu, P., Fu, A., & Zhang, Y. (2024). Uncovering access token Security flaws in Multi-User scenario of Smart Home Platforms. *IEEE Internet of Things Journal*, *1*(22), 36841–36857. Advance online publication. DOI: 10.1109/JIOT.2024.3429417

Yang, Y., Zhao, J., Li, Y., Li, X., Liu, P., & Zhang, Y. (2024). Sharing can be Threatening: Uncovering Security Flaws of RBAC Model on Smart Home Platforms. *IEEE Transactions on Dependable and Secure Computing*, •••, 1–18. DOI: 10.1109/TDSC.2024.3369634

Yassin, A. (2023). Exploring Gamification Advertising and its Role in Audience Engagement., *14*(1), 163–173. DOI: 10.21608/idj.2024.329267

Ying, X., Murugesan, R. K., Sindiramutty, S. R., Wei, G. W., Balakrishnan, S., Kumar, D., & Verma, S. (2024). Scene Text Recognition using Deep Learning Techniques. *2024 International Conference on Emerging Trends in Networks and Computer Communications (ETNCC)*, 1–9. DOI: 10.1109/ETNCC63262.2024.10767484

Youn, K., Pittman, M., & Haley, E. (2023). Using the FCB grid to understand privacy concerns in social media advertising. *International Journal of Advertising*, *43*(3), 405–433. DOI: 10.1080/02650487.2023.2242020

Yousaf, F. Z., Bredel, M., Schaller, S., & Schneider, F. (2017). NFV and SDN—Key Technology Enablers for 5G Networks. *IEEE Journal on Selected Areas in Communications*, *35*(11), 2468–2478. DOI: 10.1109/JSAC.2017.2760418

Zaheer, A., Tahir, S., Humayun, M., Almufareh, M. F., & Jhanjhi, N. Z. (2022, November). A novel machine learning technique for fake smart watches advertisement detection. In *2022 14th International Conference on Mathematics, Actuarial Science, Computer Science and Statistics (MACS)* (pp. 1–5). IEEE. DOI: 10.1109/MACS56771.2022.10023151

Zainuddin, Z., Chu, S. K. W., & Perera, C. J. (2024a). Gamification in Education. In *Sprinzer* (pp. 67–113). DOI: 10.1007/978-981-97-2219-8_3

Zainuddin, Z., Chu, S. K. W., & Perera, C. J. (2024b). Gamification Platforms for Flipped Learning Implementation. In *Springer* (pp. 167–180). DOI: 10.1007/978-981-97-2219-8_5

Zala, K., Thakkar, H. K., Jadeja, R., Singh, P., Kotecha, K., & Shukla, M. (2022). PRMS: Design and development of patients' e-healthcare records management system for privacy preservation in third party cloud platforms. *IEEE Access : Practical Innovations, Open Solutions*, *10*, 85777–85791. DOI: 10.1109/ACCESS.2022.3198094

Zarpelão, B. B., Miani, R. S., Kawakani, C. T., & de Alvarenga, S. C. (2017). A survey of intrusion detection in Internet of Things. *Journal of Network and Computer Applications*, *84*, 25–37. DOI: 10.1016/j.jnca.2017.02.009

Zen, D. N., Wibowo, D. A., & Lesmana, I. (2024). The dangers of gadget addiction on the mental health of rural adolescents in Ciamis District. www.midwifery.iocspublisher.org. https://doi.org/DOI: 10.35335/midwifery.v12i4.1694

Zhang, J. (2023, June 1). *Gamification in marketing to increase customer retention*. https://dspace.mit.edu/handle/1721.1/151418

Zhang, C., Zhang, X., & Halstead-Nussloch, R. (2014). ASSESSMENT METRICS, CHALLENGES AND STRATEGIES FOR MOBILE HEALTH APPS. *Issues in Information Systems*, *15*(2).

Zhang, H., & Sun, Q. (2023). The transformation mechanism of fitness clubs: Pricing of joint fitness courses by online platforms and well-known coaches. *Journal of Retailing and Consumer Services, 76*, 103539. DOI: 10.1016/j.jretconser.2023.103539

Zhang, J., Jiang, Q., Zhang, W., Kang, L., Lowry, P. B., & Zhang, X. (2023). Explaining the Outcomes of social Gamification: A longitudinal field experiment. *Journal of Management Information Systems, 40*(2), 401–439. DOI: 10.1080/07421222.2023.2196776

Zhang, K., Wang, Q., Qiu, L., & Wang, N. (2024). Unveiling the cost of free: How an Ad-Sponsored model affects serialized digital content creation. *Information Systems Research*. Advance online publication. DOI: 10.1287/isre.2023.0263

Zhang, L., Shao, Z., Benitez, J., & Zhang, R. (2023). How to improve user engagement and retention in mobile payment: A gamification affordance perspective. *Decision Support Systems, 168*, 113941. DOI: 10.1016/j.dss.2023.113941

Zhang, X., Kamarudin, S., Tang, Q., & Cai, Y. (2024). Within the lens of "Keep": The convergence of consumer context and brand development in Chinese fitness apps. *International Journal of Academic Research in Business & Social Sciences, 14*(1). Advance online publication. DOI: 10.6007/IJARBSS/v14-i1/20004

Zhang, Z., Chen, H., Huang, R., Zhu, L., Ma, S., Leifer, L., & Liu, W. (2024). Automated Classification of user needs for beginner user experience designers: A Kano model and text analysis approach using deep learning. *AI, 5*(1), 364–382. DOI: 10.3390/ai5010018

Zhou, W., Jia, Y., Peng, A., Zhang, Y., & Liu, P. (2018). The effect of IoT new features on security and privacy: New threats, existing solutions, and challenges yet to be solved. *IEEE Internet of Things Journal, 6*(2), 1606–1616. DOI: 10.1109/JIOT.2018.2847733

Zhou, Z., Lin, L., Wang, H., Zhou, X., Wei, G., & Wang, S. (2024, May). A Cross Domain Method for Customer Lifetime Value Prediction in Supply Chain Platform. In *Proceedings of the ACM Web Conference 2024* (pp. 4037-4046).

Zirillo, M., Aviza, S., Cruz, M., Lee, H., LaMalva, G., Galvez, A., Dragos, D., & Schmeelk, S. (2024). Examining the privacy and security of mobile refugee applications. In *Lecture notes in networks and systems* (pp. 410–427). DOI: 10.1007/978-3-031-62273-1_26

Zosh, J. M., Lytle, S. R., Golinkoff, R. M., & Hirsh-Pasek, K. (2016). Putting the education back in Educational Apps: How content and context interact to promote learning. In *Springer eBooks* (pp. 259–282). https://doi.org/DOI: 10.1007/978-3-319-45102-2_17

About The Contributors

Basheer Riskhan has completed his Ph.D. in Computer Science at Huazhong University of Science and Technology China. Dr. Basheer has an immense university/industry experience in several countries. He served as an academician and in a management position at the middle and top levels. Snowballing, he has 17 years of experience as an academician, and he served as a Lecturer, Senior lecturer, and Associate Professor. In a management position, he served as a Software Engineer, administrator, and Vice Director of foreign affairs. Currently, he is serving as Dean of the School of Computing and Informatics at Albukhary International University, Malaysia. He contributed several years as a Chief Examiner and Chief of the Evaluation Panel at the Examination Department, Ministry of Education in Information Technology, and a member of the curriculum development team at the National Institute of Education Sri Lanka. Besides academics and management, Dr. Basheer is also an excellent researcher in the domain of Data science, Big data, Cloud Counting, Machine Learning, and virtualization. He published many indexed journals in these domains and supervised students at the post-graduate level. He is also part of several active research groups and a member of research societies. Also, he provides his services as a reviewer in numerous journals and conferences. He received several awards for his research contribution, i.e., a Ph.D. scholarship award, an Outstanding Paper Presenter award, and "The Best Teacher in My Heart" award.

Khalid Hussain is working as a professor in the School of Computing and Informatics at Albukhary International University, Alor Setar Kedah Malaysia. In addition to this, he is also a member of the Senate and a member of the Board of Governor at Albukhary International University. Before this, he was working as a Professor of Cyber Security at Superior University Lahore and Campus Director at the National Superior Institute of Science and Technology (NSIST) in Islamabad; previously, he had served as Dean of the Faculty of Computer Science and Campus Director at Barani Institute of Sciences, ARID Agriculture University, for the last one and a half years. He also served at the University of Lahore as Dean of the of the Faculty of Computer Science and Information Technology for three years. Dr. Khalid has vast university and industry experience. During his tenure in the industry, he served on defense-related projects, and in recognition of his services, he has been awarded commendation certificates by multiple government agencies. He joined academia in 2008 as a full-time faculty member. In addition to his teaching role, he has been involved in numerous research projects, ICT training, seminars, and workshops, especially in the domain of cybersecurity. He helps set up a pioneering setup for information and network security certification in Pakistan. He also introduced EC Council certification under the first academia-industry partnership. That project started in a one-room office and has grown into a fully functional campus at RIPHAH University. He did his Ph.D. in Malaysia under a fully funded UTM/HHEC scholarship. Up until now, he has published 141 papers, of which 98 are ISI-indexed impact factors, 23 are in HEC-approved journals, and 20 are in IEEE and ACM conferences. Except for this, he also has 17 book chapters and three books to his credit. His cumulative impact factor is 83.032. He has completed six applied research projects in the domain of information security, funded by NESCOM and other funding agencies. Currently, he has won three projects: one from the Fundamental Research Grants Scheme (FRGS), one from the Matching Grant, and one from the Selangor Research Grant. Up until now, 21 MS and 3 Ph.D. students have completed their research theses under his supervision. Currently, he is supervising 13 MS and five Ph.D. students. In reward for his contribution towards information security, SATHA awarded him the Gold Medal in 2015.

Halawati Abd Jalil Safuan was born and raised in Kuala Lumpur, Malaysia. She received her Bachelor of Science (Honours) in Computer Science and Information Systems from University of Salford, Manchester and her Masters of Science in Technology Management from University of East London, London. She is currently the Head of Programme for Bachelor of Computer Science at School of Computing and Informatics, Albukhary International University. Her research interests are in Information Systems, Data Science and Artificial Intelligence.

***

Anzah Bashir is a dedicated academic professional currently serving as an Assistant Professor at Chandigarh University. With a strong passion for teaching and research, she has become a respected educator, inspiring her students to excel in both academics and professional pursuits. Anzah specializes in, computer science, databases, networking and is known for her innovative teaching approaches that bridge the gap between theoretical knowledge and practical applications. Her commitment to fostering a dynamic and engaging learning environment has earned her recognition among both her peers and students.

Ravikumar Ch is an accomplished professional in the field of Computer Science & Engineering, recognized for his contributions to both academia and research. He earned his B.Tech. Degree from Jawaharlal Nehru Technological University in 2004, followed by an M.Tech in Computer Science & Engineering in 2011. Recently, he successfully defended his Ph.D. in Computer Science & Engineering from Lovely Professional University in 2024, focusing on innovative approaches in Cloud Cryptography and blockchain security. With over 14 years of teaching experience and 5 years dedicated to research, Dr. Ravikumar's academic journey includes notable achievements such as receiving the Best Teacher Award from Anurag Engineering College in 2011 for his exceptional delivery of the Operating System subject. His research prowess is evidenced by his authorship of 20 Scopus indexed papers and 5 Web of Science indexed papers, alongside securing 2 patents in areas spanning cloud computing and blockchain technology. Dr. Ravikumar's expertise extends to the forefront of technological advancements, particularly in cloud computing and blockchain.

Noor Zaman Jhanjhi is a distinguished Senior Professor of Computer Science at Taylor's University, Malaysia, where he specializes in Artificial Intelligence and Cybersecurity. As the Program Director for Postgraduate Research Degree Programmes, he plays a pivotal role in shaping academic excellence and driving cutting-edge research initiatives. Globally acclaimed for his scholarly contributions, Prof. Jhanjhi has been consistently ranked among the world's top 2% research scientists (2022–2024) and stands as one of Malaysia's top computer science researchers. His exceptional work has earned him prestigious accolades, including the Outstanding Faculty Member Award (MDEC Malaysia, 2022) and the Vice Chancellor's Best Research Citations Award (Taylor's University, 2023). A prolific author and editor, Prof. Jhanjhi has published over 80 research books with leading publishers such as Springer, Elsevier, IGI Global, Bentham, IET, and Wiley, etc. amassing 1,000+ impact factor points. His mentorship spans 45 postgraduate completions, and he has examined 70+ Ph.D. and Master's theses worldwide. As an Editor-in-Chief, Associate Editor, and Editorial Board member for top-tier journals (PeerJ Computer Science, IEEE Access, CMC Computers), he advances scholarly discourse. His leadership extends to securing 40+ international research grants, underscoring his influence in innovation. A dynamic keynote speaker, Prof. Jhanjhi, has delivered 90+ invited talks and chaired major conferences. His decade-long engagement with ABET, NCAAA, and NCEAC accreditation bodies highlights his dedication to global academic standards. Combining research brilliance, academic leadership, and a passion for mentorship, Prof. Jhanjhi continues to inspire the next generation of computer scientists while shaping the future of AI and cybersecurity.

K Raghavendar is a distinguished assistant Professor in the Computer Science and Engineering department at Teegala Krishna Reddy Engineering College, Hyderabad. With 10 years of teaching experience, he has significantly contributed to the academic community through his extensive research, authoring 10 papers in various national and international journals in the field of Networks and Cloud computing.His expertise has been showcased at two international conferences, where he engaged with global peers and presented innovative ideas in the field of computer science. Born in Telangana, Dr. K Raghavendar dedication to education and research is evident in his commitment to mentoring students and fostering a collaborative learning environment.

Khalid Hafiz Mir is an accomplished academic and professional specializing in Networking and Cybersecurity. Currently serving as an Assistant Professor at Lovely Professional University, he has established himself as a dedicated educator and researcher in the field of information security. With a strong foundation in computer networks and an in-depth understanding of cyber defense mechanisms, Khalid Hafiz Mir has contributed significantly to both academic and practical advancements in cybersecurity. At Lovely Professional University, he is actively involved in shaping the next generation of cybersecurity professionals through his innovative teaching methodologies and commitment to excellence. In addition to his teaching responsibilities, Khalid Hafiz Mir is passionate about research in emerging threats in network security, ethical hacking, and data protection technologies.

Index

A

B

C

D

E

G

H

I

M

N

P

Q

R

S

U

W

www.ingramcontent.com/pod-product-compliance
Lightning Source LLC
LaVergne TN
LVHW081522300926
849161LV00006B/1266
9798369369258